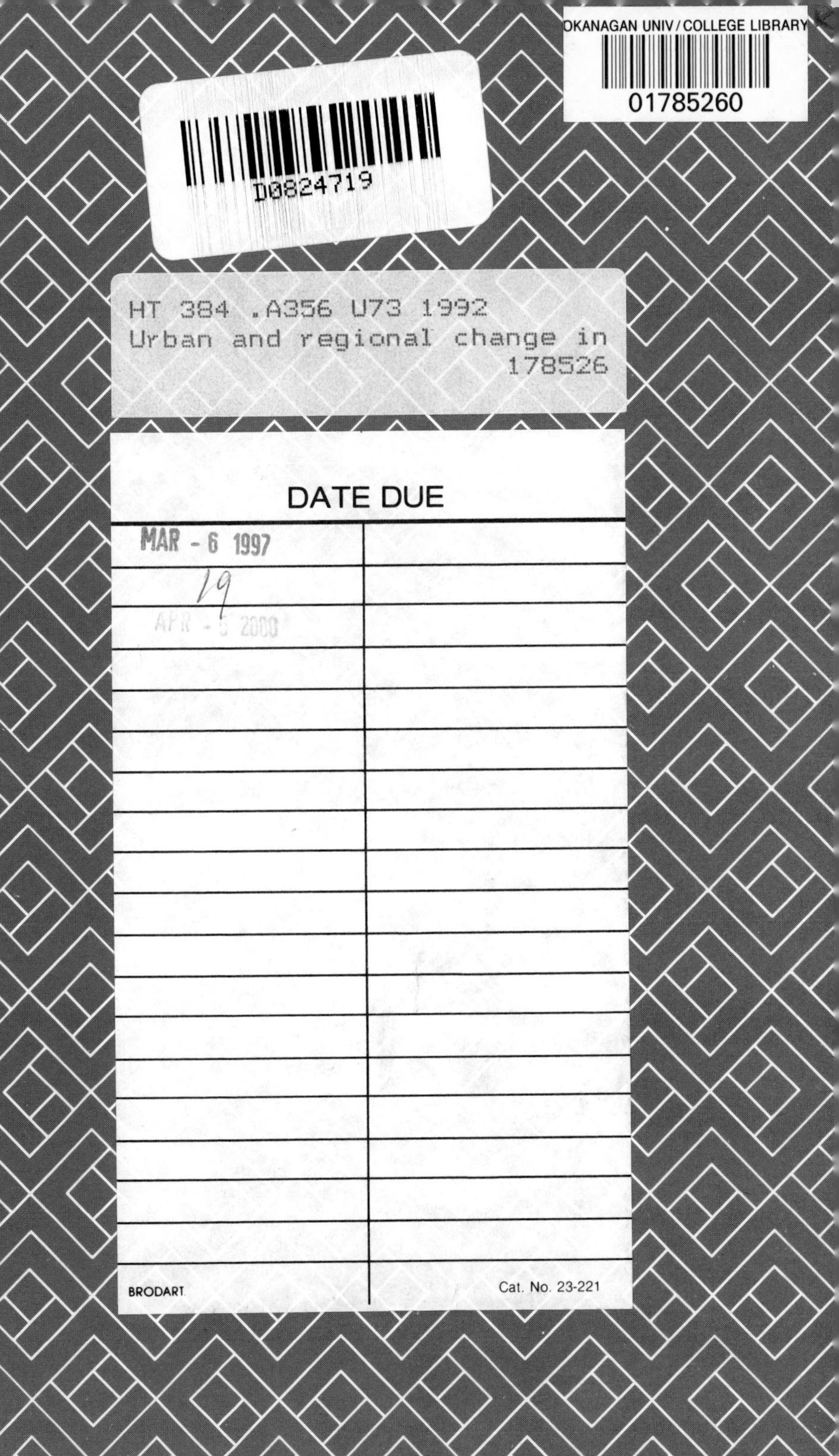

URBAN AND REGIONAL CHANGE IN SOUTHERN AFRICA

Due to the high profile of the political situation in South Africa, developments in other countries in the region tend to be overlooked. Yet the area of Southern Africa is one of the most volatile in the Third World. The central aim of this book is to examine the spatial dimensions of contemporary change in the region and to explore the likely future directions of such change.

Existing literature tends to ignore the impact of political change on the urbanization process. Moreover, it has also tended to focus on Africa as a whole or one particular state. Written by geographers and planners, *Urban and Regional Change in Southern Africa* brings a new dimension to this subject. The authors examine change in both the cities themselves and the regions in which they are located. The overall result is to set urban change in South Africa within its regional context and to contrast the impact that urbanization/political change has had in particular states, linked to each other by common concerns. The book also examines why Southern Africa should be more volatile than other developing regions, despite having undergone the same process of urbanization.

The authors have wide experience in this area and are drawn from countries that have direct or historical links with the region. The book presents a range of case study material from nation-states to households.

David Drakakis-Smith has worked in both Africa and the Far East and has published widely on development issues. He is currently Professor of Development Geography at Keele University.

URBAN AND REGIONAL CHANGE IN SOUTHERN AFRICA

Edited by
David Drakakis-Smith

London and New York

First published 1992
by Routledge
11 New Fetter Lane, London EC4P 4EE

Simultaneously published in the USA and Canada
by Routledge
a division of Routledge, Chapman and Hall Inc.
29 West 35th Street, New York, NY 10001

Typeset in Baskerville
by Pat and Anne Murphy, Highcliffe-on-Sea, Dorset
Printed in Great Britain by
Billings & Sons Limited, Worcester

British Library Cataloguing in Publication Data
Urban and regional change in Southern Africa.
I. Drakakis-Smith, D.W. (David William), *1942–*
330.968
ISBN 0–415–05441–9

Library of Congress Cataloging in Publication Data
Urban and regional change in southern Africa /
edited by David Drakakis-Smith
Includes bibliographical references and index.
ISBN 0–415–05441–9
1. Urbanization – Africa, Southern.
2. Urbanization – South Africa.
3. City planning – Africa, Southern.
4. City planning – South Africa.
5. Regional planning – Africa, Southern.
6. Regional planning – South Africa.
7. Africa, Southern – social conditions.
8. South Africa – social conditions.
I. Drakakis-Smith, D.W.
HT 384.A356U73 1992 307.76′0968–dc20
91-24981 CIP

CONTENTS

FIGURES

TABLES

CONTRIBUTORS

Keith Beavon is Professor of Geography at the University of the Witwatersrand, Johannesburg, South Africa

David Drakakis-Smith is Professor of Development Studies at the University of Keele, United Kingdom

Paul J.M. van Hoof is a staff member of the Department of Geography of Developing Countries at the University of Utrecht, the Netherlands

Anthony Lemon is a Fellow of Mansfield College, Oxford University

Carole Rakodi is Senior Lecturer in City and Regional Planning in the University of Wales College of Cardiff

David Simon is Lecturer in Geography at Royal Holloway and Bedford New College, University of London

Lovemore Zinyama is Senior Lecturer in Geography at the University of Zimbabwe

PREFACE

Most of this collection of essays was inititally presented at the First Keele Geographical Symposium held in 1988. Since then the papers have been revised and two new contributions added. The impetus behind the meeting – and the volume – was to bring together geographers who were experienced researchers in southern African affairs to comment on the spatial dimensions of change currently underway and to offer comments on the likely future directions of such changes.

Southern Africa is undoubtedly a volatile region and in recent years seems to have become even more so. Not all the changes can be viewed in a positive light: Mozambique is making increasingly desperate attempts to free itself from the nightmare of the 1980s, whilst Kenya, Tanzania and others are experiencing increasing economic pressure which in turn is generating social and political instability. On the other hand, Namibia has attained a stable independence, whilst South Africa seems to have begun to move in the right direction with the release of Nelson Mandela and the loosening of legal restraints on political organizations. For some, this suggests that a new decolonized human geography may be about to appear in that country – one which is disseminated by a growing core of Black geographers.

This is a goal that must be echoed for the region as a whole and it is to be hoped that future volumes on the geography of southern Africa, even those that emanate from British sources such as this, will contain an increasing contribution from the expanding Black geographical community in the region.

The objectives of the original conference at Keele were modest in relation to these hopes for the future. We wished to use the occasion of an academic visit by one of the contributors to this volume to draw together some of those currently undertaking research in the region to discuss their findings. Clearly, a small gathering such as this could not hope to cover a full range of countries or topics. As one might imagine within a volume of geographical essays, the dominant theme is the spatial dimension of recent change; for some this is pitched at the regional level (Lemon and Van Hoof), in others urban (Simon, Beavon, Drakakis-Smith, Rakodi) or rural (Van Hoof, Zinyama) themes dominate.

The regional focus of Anthony Lemon's chapter is very explicit. Indeed, he critically examines the new regional services councils and concludes that, although they may bring about a welcome redistribution of resources, they are deeply flawed because of the ethnically segregated local government on which they are based and in no sense can they be considered instruments of devolution.

David Simon and Keith Beavon also focus on South Africa, but more specifically on its urbanization process. Simon draws upon decolonization experiences elsewhere in the region to speculate on changes which might occur in the political economy of the South African city. Not only are racial dimensions of the urbanization process in a state of flux at present, class relations are changing too. But what does this really mean for the poor? David Simon looks closely at the situation in relation to access to land and shelter and warns that although drastic change in the built environment may not be imminent, longer-term restructuring has already begun.

In Chapter 3 Keith Beavon examines three alternative paths that urbanization may take in South Africa. The first assumes that the government will simply turn a 'blind-eye' to the darkening of grey areas in many of the larger cities; the second is constructed on the basis of a reinforced residential segregation; whilst the third is based on the repeal of such legislation. The chapter speculates on what might happen within the metropolitan and non-metropolitan areas, although cautions against anticipation of rapid or widespread changes since the built environment is expensive to reconstruct and will only change slowly. Moreover, even if the legislative framework of apartheid is dismantled the continued practice of racial discrimination may well perpetuate inequality in access to resources for a considerable time.

Chapter 4 shifts the focus of attention to Zimbabwe, which many consider may provide a model of what South Africa may become if the African National Congress (ANC) one day assumes the responsibilities of government. In this chapter David Drakakis-Smith examines the impact of ten years on socialism on the welfare of the urban poor, particularly with respect to their access to housing and food. He concludes that as yet few signs of socialism are evident in the urban social welfare programmes and that given government rhetoric the results have been more disappointing than in other sectors of Zimbabwean development.

Carole Rakodi takes up similar themes in her review of the planning and management of urban development in Zambia, Tanzania, and Zimbabwe. Her chapter focuses on land markets, housing, and household survival strategies in the context of the dynamics of the planning system. The chapter clearly identifies the importance of the relationship between central and local government and the extent to which this varies in the three countries, with important consequences for the planning process.

The final two chapters direct attention more towards rural development, first in Botswana and second in Zimbabwe. Paul van Hoof set his analysis of rural development in the context of regional planning and the way in which this is structured within the national planning framework, the regional difference in resource availability, and the degree of political decentralization. In this context there are clear links with earlier chapters by Lemon and Rakodi. In similar vein he concludes that real change at the local level will only follow effective devolution of decision-making.

Lovemore Zinyama, in contrast, focuses on the transformation of the small-scale farming sector in Zimbabwe through the adoption of new yield-increasing technologies. By means of a series of case studies, he paints an overall picture of rising production levels and rightly draws attention to the success of cooperative efforts in expanding crop output. However, access to technology is not the only factor which has proved to be important, farmers must possess adequate land, labour, and other resources before full benefits can be realized.

The task of drawing these varied contributions together from their origins on a snowy winter's day in Keele has fallen on many individuals other than the editor and I would like to thank Pauline Jones, May Bowers, Andrew Lawrence, and Marylyn Beech for their invaluable assistance.

David Drakakis-Smith
University of Keele

ABBREVIATIONS

ACTSTOP	Action Committee to Stop Evictions
AFC	Agricultural Finance Corporation
ANC	African National Congress
AP	annual plans
ARDP	Accelerated Rural Development Programme
ASH	aided self-help housing
BDP	Botswana Democratic Party
BLAs	Black local authorities
BNF	Botsana National Front
BPP	Botswana People's Party
CAF	Central Africa Federation
CBD	central business district
CFDA	Communal First Development Area
CMB	Cotton Marketing Board
CP	Conservative Party
CPO	council planning officer
CTCC	Cape Town City Council
CTP	Cape Town and the Peninsula
DA	district administration
DC	district commissioner
DDC	District Development Committee
DDPs	district development plans
DLUPU	District Land Use Planning Unit
DO	district officer
DOD	district officer (development)
DPC	District Plans Committee
DTRP	Department of Town and Regional Planning
ESCOM	Electricity Supply Commission
FAP	Financial Assistance Programme

GGP	Gross Geographic Product
GMB	Grain Marketing Board
GST	general sales tax
HYV	high-yielding variety
JEA	Joint Executive Authority
JMCs	joint management centres
MDM	Mass Democratic Movement
MFDP	Ministry of Finance and Development Planning
MJSBs	metropolitan joint services boards
MLGL	Ministry of Local Government and Lands
NDP	National Development Plan
PEU	Port Elizabeth–Uitenhage
PLAs	primary local authorities
PWV	Pretoria–Witwatersrand–Vereeniging
RDC	Rural Development Council
RDU	Rural Development Unit
RSA	Republic of South Africa
RSC	regional services council
SADCC	South African Development Cooperation Conference
SAL	Structural Adjustment Lending
SIDA	Swedish International Development Agency
TGLP	Tribal Grazing Land Policy
THB	Tanzania Housing Bank
TTLs	Tribal Trust Lands
UBCs	urban Bantu councils
UDCORP	urban development corporation
UDF	United Democratic Front
UDI	Unilateral Declaration of Independence
ULCC	ultra-low-cost core
ULGS	Unified Local Government Service
UNICEF	United Nations Children's Fund
USAID	United States Agency for International Development
VDCs	village development committees
ZANU-PF	Zimbabwe African National Union-Patriotic Front

1

RESTRUCTURING THE LOCAL STATE IN SOUTH AFRICA

Regional services councils, redistribution and legitimacy

Anthony Lemon[1]

Local government in South Africa is widely regarded as a sensitive and highly politicized issue which is critically important to the resolution or intensification of the wider South African crisis. Extra-institutional opposition groups have identified local communities as vital units around which their struggle should be organized (Atkinson and Heymans 1988: 155). State planners, on the other hand, see the reconstitution of local government as the basis for the subsequent reorganization of second-tier (regional) and first-tier (central) structures. They are starting at the bottom in this way because they fear that to start by reconstituting central government institutions would provoke the mobilization of significant right-wing White opposition to 'power-sharing', as well as head-on confrontation with extra-institutional forces which have already proved their ability to undermine the reform process through boycott action (Swilling 1988).

This chapter focuses on the new regional services councils (RSCs) which are indirectly elected bodies covering the area of several local authorities. As their name implies, their rationale is theoretically the need for bodies to provide certain services more efficiently at a larger scale than that of local authorities. However, they are also charged with the responsibility of upgrading infrastructure in those parts of their areas where the need is greatest, and it is this function which makes the central plank of state attempts to legitimize third-tier local government structures. As such, they are destined to play a critical role in the success or otherwise of the whole constitutional reform strategy. The implementation and functioning of RSCs can only be understood and evaluated in relation to that strategy, and in the context of the development and financing of Black local government under apartheid. It is therefore to these issues that we now turn.

ANTHONY LEMON

THE DEVELOPMENT OF APARTHEID STRUCTURES IN LOCAL GOVERNMENT

For most of South Africa's history, local government has been in White hands. For a long time this was largely unaffected by the consolidation of ethnic residential segregation in urban areas, first with respect to Africans in terms of the Natives (Urban Areas) Act of 1923, and subsequently with respect to Whites, Coloureds, and Indians[2] in terms of the Group Areas Acts of 1950 and 1966 which have effected the transition from 'segregation cities' to 'apartheid cities' (Davies 1986). Although spatially segregated, people of all four of these officially recognized population groups continued to fall under the direct control of established local councils which were representative of a largely exclusively White electorate. Where Coloureds and Indians have been able to qualify for the franchise, as in Cape Town, they were eventually removed from the municipal rolls in 1971.

The development of segregated local government bodies for Africans has followed a separate path from that for Coloureds and Indians, but in each case it has accelerated rapidly in the 1980s. For Coloureds, Indians, and Whites local government was deemed to be an 'own affair' in terms of the 1983 Constitution, but for Africans, who were excluded from the new tricameral parliament, it was treated as a 'general affair' under the direction of the Ministry of Co-operation and Development and regional administration (later development) boards.

Local government for Coloureds and Indians

Legislation providing for the staged evolution of representative Coloured and Indian municipal councils was embodied in the Group Areas Amendment Act of 1962. The process was to begin with the appointment (by provincial administrators) of committees intended to function in a purely consultative capacity in relation to White local authorities which retained administrative control over their areas. These committees were subsequently to be transformed into partly, and later wholly, elected management committees with advisory functions in relation to a prescribed range of local issues. By 1975 there were ninety-seven Coloured management committees – of which eighty-one had elected as well as nominated members – and four Coloured local affairs committees in Natal, of which two were fully elective. Most Indian townships had only nominated local officers, management or consultative committees; Isipingo, near Durban, became a fully-fledged Indian local authority in 1975, and town boards functioned at Verulam and Umzinto.

By 1980 there was some increase in democracy if not in power: all 171 Indian and Coloured management committees and six local affairs committees in Natal were fully elective, but they remained advisory. Consultation by White local authorities was frequently inadequate, and there was growing dissatisfaction and frustration with the system on the part of its Coloured and Indian members. They increasingly demanded the abolition of management and local affairs committees and the establishment of direct representation on a non-racial basis on city and town councils. One city council, Cape Town, expressed its wish for such representation by all ratepayers in 1985. None of the city's Coloured management committees was in a position to be promoted to full local authority status in terms of the criteria set out by the provincial administrator; these included the stipulation that size and level of development should allow for independent progress and financial autonomy (Todes *et al.* 1986: 6).

The failures of the system were noted and analysed by several official reports, beginning with that of the Theron Commission (South Africa 1976), which identified problems of both legitimacy and financial organizational viability. The problem of financial viability in Coloured areas was underlined by the Yeld Committee (South Africa 1978a), whilst in the same year the Schlebusch Committee pointed to the political failure of the system (South Africa 1978b). The Slatter Committee reached similar conclusions in 1979 in relation to four Indian areas in the Transvaal and Natal (South Africa 1979).

It was the report of the Browne Committee in 1980 which sowed the seeds from which RSCs were eventually to grow. Ironically, its appointment arose from the problems of White local authorities faced by a steadily eroding rate base caused by inflation and slow economic growth (Leon 1987: 176). Browne concluded, however, that while the revenue bases of White authorities might be shrinking, the greatest needs lay elsewhere, and that the financial problems of management and local affairs committees could only be met by a system of transfer payments from White local authorities (South Africa 1980). This was not surprisingly opposed by the White United Municipal Executive, and would undoubtedly have aroused strong opposition from White ratepayers and loss of electoral support by the government.

This thorny question was referred to yet another committee, the

Croeser Working Group, which accepted the findings of the Browne Committee, but instead of direct transfer payments proposed what amounted to a system of fiscal laundering through which payments from the affluent White commercial and industrial sectors could be channelled to those areas most in need of infrastructural and environmental upliftment (South Africa 1982). Two of the three taxes specifically recommended by Croeser – a regional services levy and a regional establishment levy – were later to be introduced as the basis of funding regional services councils.

Meanwhile the government attempted to address the political deficiencies of the management committee system. In November 1984 it published regulations designed to improve communication between White local authorities and management committees by requiring the establishment of liaison structures. These were aimed at both conservative White authorities which had been unwilling to engage in serious consultation hitherto, and relatively liberal authorities such as Cape Town which rejected the principle of ethnic separation in local government inherent in the management committee system.

This was followed by the Local Government Affairs Act of 1985, section 17b of which provides for the transfer of specified functions from White local authorities to management committees. This potential creation of more than one decision-making body within a single local jurisdictional area is confusing and potentially conflict-ridden. In the words of the Town Clerk of Cape Town:

> It can only work with benevolent co-operative attitudes otherwise it could lead to 'we–they' division of the unity of jurisdiction and administration . . . normally found in a local authority, without anything like full autonomy. It can only be an interim step towards something else.
>
> (Evans 1985: 14)

That 'something else' could, in terms of official thinking, only be full municipal autonomy for Coloured and Indian areas. The financial viability and political legitimacy which had hitherto proved so elusive would, it was hoped, flow from the implementation of RSCs. But most management committees apart from Athlone (Cape Town) and Ennerdale (southern Transvaal) are unwilling to accept full autonomy, preferring to hold to the ideal of non-racial, unified local government.

Local government for 'non-homeland' Africans

Prior to 1977 the only official representation permitted to urban Africans was through advisory boards and urban Bantu councils (UBCs), neither of which significantly influenced the decisions of White municipalities. They remained equally powerless after 1972, when African townships became the responsibility of central government in the shape of newly created regional administration boards. In the late 1960s and early 1970s even such token representation was felt to be inappropriate by state officials, who favoured the replacement of advisory boards and UBCs with a system of 'homeland government' representatives in urban areas, in conformity with the Verwoerdian myth that all Africans were but 'temporarily sojourners' in 'White' urban areas.

It took the widespread township unrest of 1976, the year of the Soweto riots, to persuade the government of the unreality of such an approach. Its response was the creation in 1977 of community councils, to which the Minister of Co-operation and Development was empowered to transfer certain functions, including the allocation and administration of accommodation, from administration boards. The greatest transfer of functions generally occurred in the larger urban areas, but it was also in these areas that polls were lowest. In Soweto, for instance, the steady decline in the percentage poll in UBC and community council elections from 32 per cent in 1968 to 14 per cent in 1974 and only 6 per cent in 1978 clearly indicated rejection of officially sanctioned structures (Bekker and Humphries 1985: 105). The image of the councils was not improved by their ambiguous relationship with administration boards; in particular their control of housing allocation necessarily involved the councils in influx control.

The government chose to attribute the failure of community councils to their lack of autonomy rather than to the fiscal unviability and perceived lack of political legitimacy around which literally hundreds of community organizations were mobilizing. Its response was contained in the Black Local Authorities Act of 1982 which created two new structures, town and village Councils, and also provided for advisory committees in smaller urban areas. The vesting of powers in village councils is still subject to ministerial discretion, but town councils are officially regarded as fully independent Black local authorities (BLAs). However, they remain subject to wide-ranging powers of ministerial intervention which far exceed those affecting White local authorities. Failure to agree 'adequate charges' for municipal services is a particularly sensitive ground for intervention, given the politicization of such charges in recent years.

The Black Local Authorities Act again involves the councils in the administration of *de facto* influx control, this time more explicitly in the exercise of their scheduled powers; these include the approval of building plans, the removal or demolition of unauthorized structures, and the prevention of unlawful occupation of housing. The formal abolition of racially discriminatory influx control in 1986 has done little to diminish the impact of these functions on would-be African urbanites. Meanwhile BLAs continued to rely on the administration boards for seconded administrative staff, whose loyalties were questionable; relations between the BLAs and the boards were strained, while relations between the boards and the Department of Co-operation and Development, which had to mediate between the boards and the BLAs, were also problematic (Humphries 1988: 108; Bekker and Humphries 1985).

The government attempted to deal with these problems in 1985 by transferring responsibility for the monitoring of the Black Local Authorities Act to the Ministry for Constitutional Development and Planning, a move supposed to increase the legitimacy of BLAs. In July 1986 the development boards (as the administation boards had been rechristened) were abolished, the abolition of influx control having removed a major rationale for their existence. BLAs were thus freed from a relationship which they had resented, and the Department of Constitutional Development and Planning (DCDP) now faced the task of improving their legitimacy.

However, such bureaucratic reordering did nothing in itself to improve the financial position of the BLAs. The state chose to equate municipal independence with financial independence, but the expectation of financial self-sufficiency was inimical to attempts by BLAs to gain legitimacy. Group areas and related legislation has prevented African townships from attracting industries and minimized the level of commercial development, thus depriving the townships of a potentially important source of revenue. Given the poverty of most of their residents, housing is mainly sub-economic rental property; the Black Communities Development Act of 1984 does provide for the private sale of property in the townships, but financial constraints and political uncertainties have minimized the effect of this change in providing BLAs with a property tax base from which to obtain revenue. Likewise, income from services such as water, electricity, and refuse removal cannot provide a sufficient revenue base for the upgrading of infrastructure and social facilities, which is certainly a necessary if not a sufficient condition for improving legitimacy. The privatization of beer and sorghum halls, the profits of which had provided the administration boards with resources for township development, has deprived the BLAs of even this traditional revenue source.

Many BLAs felt that their only option was to increase rents, often substantially. But this led to resistance from communities who could ill afford such increases, especially at a time of economic recession and increasing unemployment. It was the decision of the southern Transvaal local authority of Sebokeng to increase rents in 1984 which triggered off the wave of unrest which quickly spread to much of the country and continued until brought under control by police and army action in 1986. Rent boycotts soon became a key form of resistance to the state, and by September 1986 they affected 25 African townships in the Transvaal, 7 in the Orange Free State, 19 in the eastern Cape and 3 elsewhere (Swilling 1988: 200).

The combination of financial and political pressures, including the growing physical threat to councillors branded as 'collaborators', caused the complete collapse of many BLAs, especially in the eastern Cape and the Pretoria–Witwatersrand–Vereeniging area. This forced the DCDP to appoint administrators to oversee the continued day-to-day administration of the townships concerned.

The transfer of responsibility for BLAs to the DCDP coincided with a change of emphasis towards the representation of Africans in 'general affairs' structures. BLAs became the first institutions to be affected by this policy, as the DCDP decided to use the proposed RSCs as the means to improve the legitimacy of BLAs. Whereas the 1984 draft bill on RSCs had applied only to White, Coloured and Indian local authorities, provision was made for the representation of BLAs in the version placed before parliament in 1985, which was duly enacted. This is clearly a critical change in the RSC concept, and one which has potentially far-reaching implications, both for constitutional change at higher levels and, more immediately, for the redistribution of wealth within local areas (see pp. 24–25).

In making this change the DCDP signalled its realization that financially self-sufficient BLAs could not work. However, just as the failure of community councils was attributed in 1982 to lack of autonomy rather than lack of fiscal viability, failure of BLAs was now attributed solely to financial problems rather than to their fundamentally unacceptable racially defined nature. This is not untypical of the way in which the South African state has repeatedly grasped only that part of the nettle which it is compelled to see at a particular juncture.

REGIONAL SERVICES COUNCILS AND THE 1983 CONSTITUTION

South Africa's 1983 Constitution provides for a tricameral parliament consisting of the House of Assembly (White), the House of Representatives (Coloured), and the House of Delegates (Indian). Each house is responsible for what the Constitution deems to be the 'own affairs' of its population group (see Figure 1.1); these include – with qualifications – social welfare, education, art, culture, recreation, health, housing, community development, local government, agriculture, and water affairs. There is a ministers' council for each population group, and executive authority for 'own affairs' is vested in the State President acting on the advice of the appropriate ministers' council. A structure separate from the provincial councils is planned, with regional offices linking central government to local government at the 'own affairs' level (see Figure 1.1). Executive authority for other matters, known as 'general affairs', is vested in the State President and the Cabinet, whose ministers he appoints to administer departments of state.

A new and much more centralized system of provincial government has also been introduced, and came into operation on 1 July 1986. It replaces the former elected White provincial councils with provincial administors appointed by central government and multiracial nominated provincial executive committees. That, in the Cape for instance, includes three Whites who are all ex-members of the provincial council; two Coloureds, a teacher, and a director of housing for the Port Elizabeth municipality; and one African businessman who is also mayor of this township in Kimberley.

Provision is made for a representative function in the form of a standing parliamentary committee for each province, which further underlines the centralization which has occurred. As there are no Africans in parliament, the standing committees lack African members. Their composition is, oddly, proportionate to the representation of all parties in the three houses of parliament, regardless of the cross-section of parliamentary representation in question. This makes the committees bizarrely unrepresentative: the parliamentary standing committee for Natal has two Conservative MPs although the

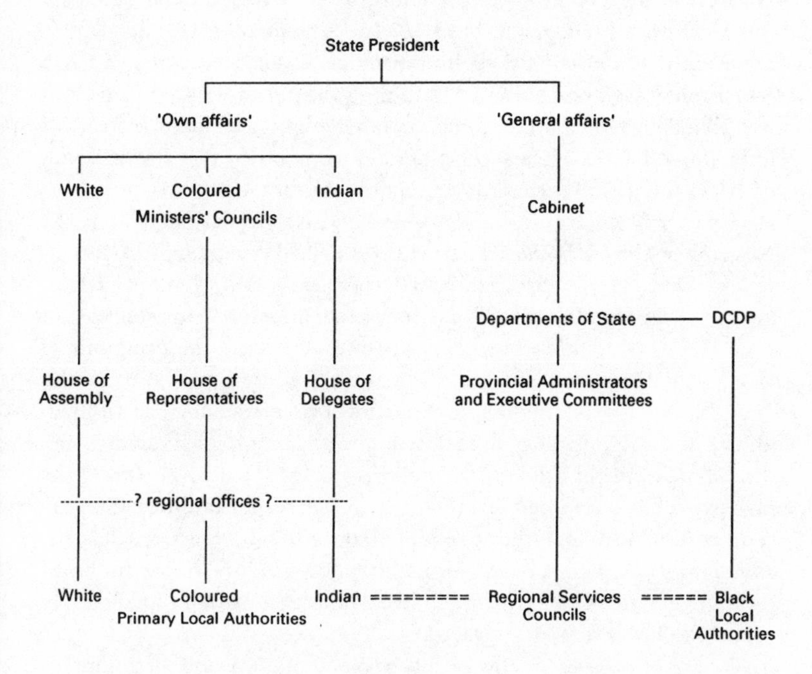

Figure 1.1 Constitutional structure of 'own' and 'general' affairs

province returned no candidates for this far-Right party, whilst the Orange Free State committee includes five Indian MPs in a province which until recently prevented Indians from living there!

These new provincial administrations are seen as facilitators of the official concept of 'power-sharing' at the second tier of government (Heymans 1988: 43). As such they are responsible for 'general affairs', and their character is consonant with what appears to be a policy of providing multi-racial representation only on non-elective executive or functional bodies entrusted with the execution or implementation of 'general affairs' (Olivier 1987: 28). The same is true of RSCs themselves, the Co-ordinating Council for Local Government and its action committee, and the recently created Local Government Training Board. Elective bodies with some degree of legislative power are, on the other hand, based firmly on the principle of racial separation.

There are, however, indications that this may shortly change. The Minister of Constitutional Development and Planning, Mr Chris

Heunis, hinted at the reintroduction of elected provincial councils, this time on a multi-racial basis (*Beeld*, 24 September 1987). This is apparently supported by all four provincial administrators. Such a reform may well be combined with the splitting of the vast Cape and Transvaal provinces into three, making eight provinces instead of four altogether. Such a second-tier reform would, it is felt in some quarters, be more efficient and fit more logically with the regionalization patterns emerging under the RSC system (*Business Day*, 12 November 1987; *Diamond Fields Advertiser*, 13 November 1987).

For Africans, no distinction between 'own' and 'general' affairs has yet been made. All Black affairs fall under the jurisdiction of central government departments of state, including the functions of BLAs which we have already seen to be the responsibility of the DCDP. Coloured, Indian and White local government is, as an 'own affair', the responsibility of the local government departments of the three houses in the tricameral parliament. In practice, however, the provinces have retained many of their old responsibilities thus far (and all four provinces have retained administrative sections responsible for local government), although the 'own affairs' departments of local government have assumed the major responsibility for housing matters.

A similar confusion affects the implementation and functions of RSCs themselves. Responsibility for their implementation has been assumed by the provincial administrations on the grounds of their being 'general affairs' bodies. In practice, as will be seen, RSCs have very direct effects on the 'own affairs' responsibilities of White, African, Coloured and Indian 'primary local authorities' (PLAs) at the third tier. Officially RSCs are regarded as horizontal extensions of the PLAs, but in practice they constitute an extra tier of government which, although only indirectly elected, has mandatory executive and taxing powers over PLAs (Cameron 1988: 55).

THE STRUCTURE OF REGIONAL SERVICES COUNCILS

The specific formulation of the Regional Services Councils Act of 1985 can be traced to the Joint Report of the Committee for Economic Affairs and the Constitutional Committee of the President's Council in 1982, which recommended the introduction of a regional system for the provision of 'hard' or 'bulk' services in the bigger urban areas and some rural areas. After referral, first to the Co-ordinating Council for Local Government Affairs and subsequently to a committee (the

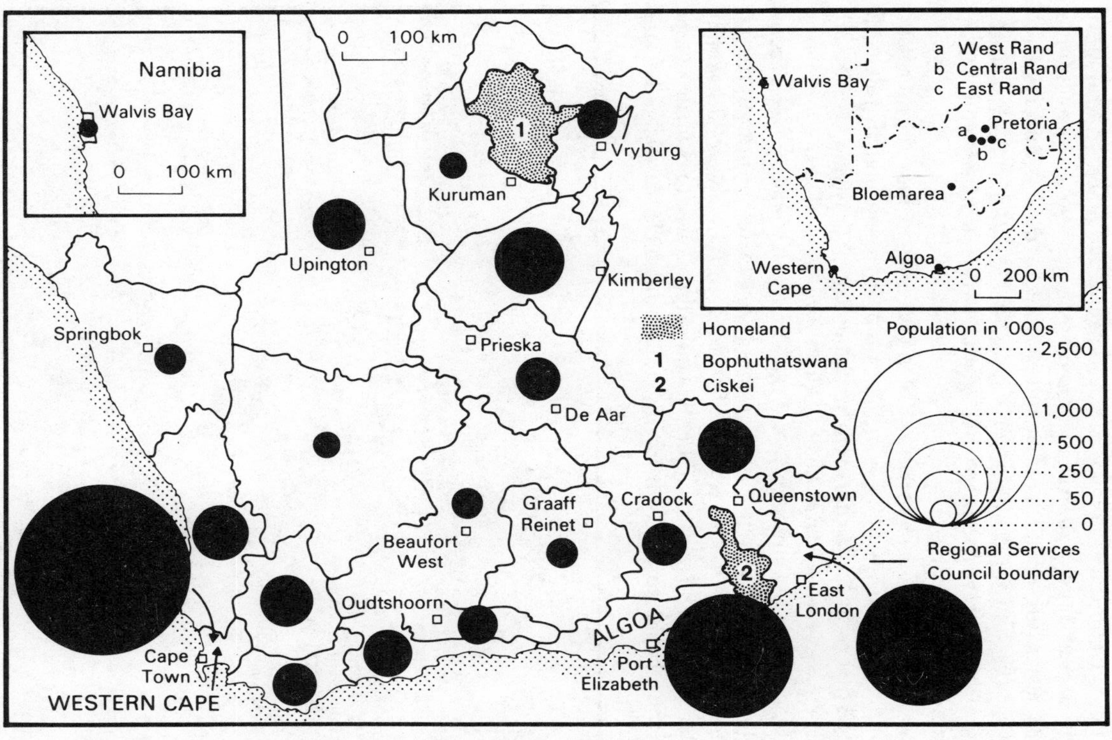

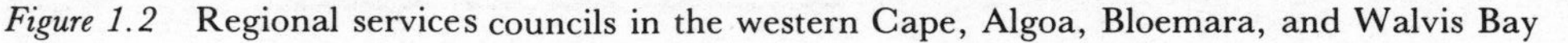

Figure 1.2 Regional services councils in the western Cape, Algoa, Bloemara, and Walvis Bay

Louw Committee), a bill was placed before parliament in May 1984. It caused much controversy, mainly because of the exclusion of Africans, the ethnic basis of representation, the system of indirect election, and the proposed revenue sources. A revised bill allowing for the representation of BLAs was passed in 1985 and further amended in 1986.

The formation of the first RSCs was delayed for a further year, mainly due to problems with their proposed financing. In July 1987 the first eight RSCs came into being in Central, East and West Rand, Pretoria, the western Cape, Algoa (the Port Elizabeth region), Bloemarea (Bloemfontein), and the Walvis Bay enclave (see Figures 1.2 and 1.3). Since then the Transvaal has made the fastest progress in establishing RSCs: ten out of a proposed twelve were functioning by March 1988, and the other two quickly followed. Six further Cape RSCs began to function in July 1989, and the final twelve became

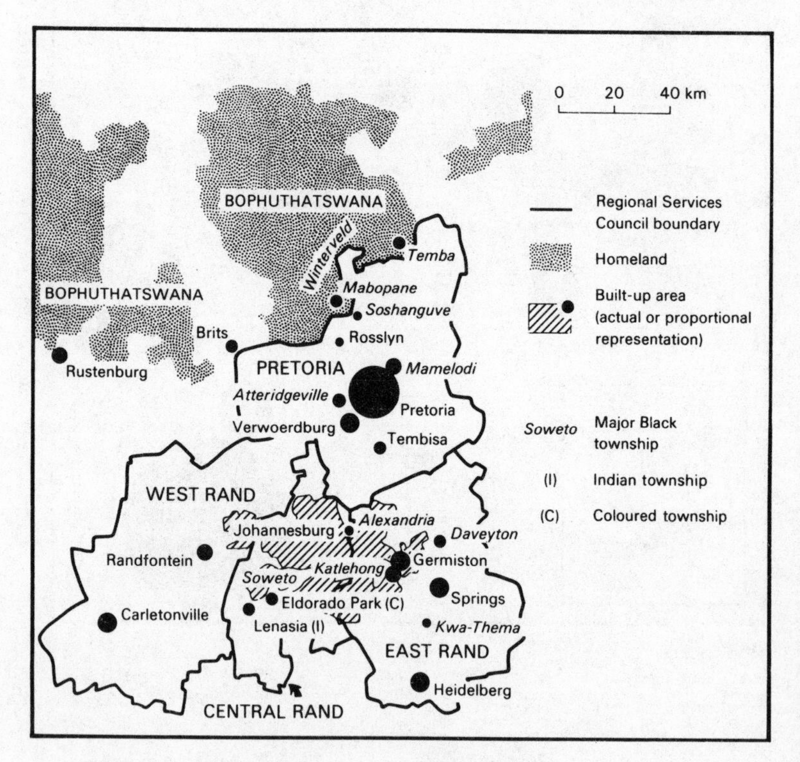

Figure 1.3 Regional services councils in Pretoria, and in Central, East, and West Rand

operative in 1990. In the Orange Free State the Goldfields RSC came into being in October 1989, but the eastern and northern RSCs exist only on paper at the time of writing. In Natal, implementation has been delayed by the opposition of Chief Mangosuthu Buthelezi and his KwaZulu 'homeland' administration. Initial proposals for four RSCs which excluded KwaZulu territory were published in March 1988, but these were superseded by new proposals for joint authorities in 1989.

According to Chris Heunis (*Die Vaderland*, 28 November 1985) the officially stated aims of RSCs are threefold:

1 To increase cost-efficiency and effectiveness in hard service provision.
2 To introduce multi-racial decision-making at the third tier.
3 To promote infrastructural development, from new revenue sources, in African, Coloured, and Indian townships which are commonly accepted as the areas of 'greatest need', to which the Act specifically refers.

Schedule 2 of the RSC Act lists twenty-two services which RSCs may be required by the provincial administrator to provide. Among the more important are the bulk supply of water and electricity, sewerage, land use and transport planning, roads and passenger transport services. The last two are catch-all clauses: 'the establishment and maintenance of other infrastructural services and facilities' and 'other regional functions'. RSCs therefore undertake virtually anything they choose, subject to the consent of the provincial administrator. In the event of their taking over all twenty of the specific functions listed, elected PLAs would lose most of their more important functions apart from housing.

RSCs will have two new sources of revenue in addition to user charges for the services provided. Both are levies on business rather than on householders, thus avoiding the adverse political impact of the transfer payments proposed by the Browne Committee. The maximum levels of both levies are set by the Minister of Finance. An establishment levy based on turnover has initially been set at 0.1 per cent and a services levy based on an employer's payroll has been set at 0.25 per cent. Of total levy revenue, 5 per cent must be placed in a local government training fund, a provision designed to provide training for those councillors who have little or no previous experience.

Formally at least, RSCs break new ground in the constitutional history of South Africa as multi-racial decision-making bodies. RSC members are nominated by all constituent PLAs as well as other elected local bodies such as management committees where no PLA exists. However, representation is based not on population but on the amount which a PLA or other body pays for the services provided, with one councillor for each 10 per cent of the total voting power or part thereof. Table 1.1 shows the practical outcome of this in the Central Witwatersrand in 1989–90.

Table 1.1 Determination of voting power on the Central Witwatersrand Regional Services Council

Local body	*Expenditure** *(R)*	*Voting power* *(%)*	*Number of representatives*
Alexandra	3,986,820	0.83	1
Diepmeadow	30,886,376	6.43	1
Dobsonville	4,751,670	0.99	1
Ennerdale	4,651,233	0.97	1
Johannesburg:			
Indian	5,024,014	1.04	1
Coloured	7,410,808	1.54	1
White	193,510,614	40.28	5
Lenasia	6,385,962	1.33	1
Marlboro Gardens	24,942	0.01	1
Randburg	40,868,948	8.50	1
Roodeport:			
Coloured	1,207,140	0.25	1
White	73,695,657	15.34	2
Sandton	25,067,892	5.22	1
Soweto	82,968,966	17.27	2
Total	480,441,042	100.00	20

The bias towards wealthier White areas which this implies is somewhat mitigated by three further provisions:

1 That central business districts and industrial areas are excluded from the calculation.
2 That no one PLA may have more than 50 per cent of the voting power.
3 That decisions must be taken by consensus or by a two-thirds majority.

Thus no single authority can dominate decision-making, although two or more White PLAs will usually command a two-thirds majority between them. The allocation of voting strength is to be reviewed annually, and it is implicit that the upgrading of services in African, Coloured and Indian areas will gradually increase their consumption and hence their voting power. Thus the voting power of Soweto in the Central Witwatersrand RSC increased from 12.76 per cent in 1987/8 to 16.54 per cent in 1988/9 and 17.27 per cent in 1989/90. Needless to say, such 'progress' is wholly insignificant in the context of Black aspirations.

The provincial administrator has a pivotal role in relation to the functioning of RSCs. He has the responsibility of deciding on its areal extent after considering the report of a demarcation board, and of identifying which central business districts and industrial areas should be excluded for the purposes of calculating voting strength. He decides what functions the RSC should perform. He also appoints the RSC chairman who is a full-time salaried official.

There is provision in the RSC Act for any PLA or other local body which is unhappy with an RSC decision to ask the council to debate the issue afresh. If still dissatisfied, an appeal can be made. The appeal board consists of the provincial administrator (or the Minister of Finance in respect of financial matters), the Minister of Constitutional Development and Planning (in his capacity as minister in charge of Black local government), and the three 'own affairs' ministers of local government.

The Act provides for the inclusion within RSCs of areas belonging to 'homelands' or 'national states', with the agreement of the relevant authorities. The voting strength allocated to local bodies in these areas is to be determined by an agreement which is subject to approval by the provincial administrator and (because of the officially self-governing or independent status of these territories) the Minister of Foreign Affairs.

THEORETICAL CRITICISMS OF REGIONAL SERVICES COUNCILS

RSCs attracted much wide-ranging and penetrating criticism well before their actual establishment (Todes and Watson 1985a, 1986a, 1986b; Todes *et al.* 1986; Lear and Winter 1985; de Tolly 1986; Bekker *et al.* 1986; South African Institute of Race Relations 1987a). The most fundamental criticisms which surface repeatedly in one form or another may all be subsumed under the broad heading of legitimacy. First, however, we will briefly review criticisms concerning the efficiency and financing of RSCs, and the centralization implicit in their establishment and operation.

Efficiency

Local government reform in many countries has involved the amalgamation of local authorities and the formation of metropolitan tiers of government. Invariably this is done in the name of efficiency; it is argued that fragmentation and duplication of service provision can be eliminated, and that economies of scale can be realized. In South Africa such arguments were used by committees of the President's Council regarding the provision of bulk services (South Africa 1982: par. 6.23), and appear to be the official justification for the servicing function of RSCs.

The evidence for economies of scale is inconclusive (Todes and Watson 1985b). In a number of metropolitan reorganizations 'economy' and 'efficiency' results have been disappointing (Jones and O'Donnell 1980). Different services will in any case have different optimum scales of operation, which casts doubt on the efficiency of any one scale of local government unit for the provision of a wide range of services. Nor is it usually practicable to make all units of a given tier similar in population size; thus existing and proposed RSCs in the Cape province range from six areas of less than 100,000 people to the Western Cape RSC which has a population of 2.2m. (see Figure 1.2).

Fragmentation and duplication are genuine problems, but amalgamation of local authorities is not the only, or necessarily the best, way to solve them. In conurbations some form of city-wide coordination may be necessary to overcome fragmentation, but this need not entail the existence of a metropolitan body to provide the services. In South Africa, as in the United States, many towns already enter into service agreements with their smaller neighbours. It is also common for municipalities to enter into contractual arrangements with private firms for such services as refuse collection and park maintenance; such firms can pass on the economies of scale in their own large operations. Parastatals such as ESCOM (the Electricity Supply Commission) and the Rand Water Board also provide bulk services to many local authorities; the introduction of an intermediate body, the RSC, to buy and resell these services to constituent local authorities hardly constitutes improved efficiency.

In metropolitan areas, some sort of coordinating mechanism is clearly necessary for planning purposes, especially under the conditions of rapid urbanization and growth which prevail in all major South African cities. To satisfy this need, either metropolitan authorities must have strong representation on regional or provincial authorities, or some form of coordination must exist between metropolitan and peri-urban and rural government structures (Todes and Watson 1985b: 43). The planning function of RSCs does not meet these needs because it is confined to the implementation of plans formulated at national level (Todes *et al.* 1986: 34–5). This is true both of policies regarding the location of industrial and urban development, and also of more detailed guide plans for metropolitan areas which are formulated by guide plan committees appointed by the Minister of Constitutional Development and Planning. The latter are required to formulate their plans within the framework of national policies, and RSCs are required to ensure that land use proposals are consistent with the guide plan. They therefore have no power to extend industrial development beyond the limits laid down in the guide plan, or to make decisions relating to levels of urbanization and city growth.

Financing

The RSC levies have been criticized on several grounds. There are fears that they will be inflationary as the costs are passed on to consumers, and that the payroll tax will hinder much needed job creation. The annual cost to industry has been variously estimated at R635m (*Financial Mail*, 20 March 1987), R692m (Assocom 1987), and officially at R800m (Porteous 1987: 9). This is not large in relation to the total corporate tax burden of R4.5b in 1984 (Porteous 1987: 10), but administrative costs to business and industry (especially of the turnover tax) may also double the levy burden (Assocom 1987). There is also an understandable fear that, given the immensity of the upgrading task in many RSCs, the maximum permitted levels of taxation will quickly rise, as has happened during the 1980s after the introduction of general sales tax (GST). By mid-1989 only Pretoria had applied for an increase.

There is some compensation for these new tax burdens. Business and industry are relieved of the cost of trade licences, transport levies, contributions to development boards and, in the Cape, divisional council rates (see pp. 23–24). In the Central Witwatersrand RSC these gains were expected to offset nearly half the cost of levy payments, excluding the administrative costs involved.

Undesirable geographical consequences of the levies have also been suggested. In order to avoid paying turnover tax several times over, vertically integrated concerns could seek to concentrate their activities in one RSC area, contrary to official policies of deconcentration (Ashcroft 1987). The fear that businesses could relocate outside RSC areas seems baseless, as these will soon cover the whole country outside Natal, nor is there is any evidence as yet of differential rates; surprisingly, not all RSCs have chosen to levy the maximum permitted rates. The redistributive effects of the levies have also been questioned on the grounds that they will be passed on to consumers and workers – both of whom are mainly Black – in terms of higher prices, lower wage increases, and unemployment (Wolff 1987).

Most fundamental is the criticism that the state has, by using the RSCs as agents of redistribution, rejected other means at its disposal, above all through the channelling of national revenues to subsidize local needs. This is inherent in, for example, the British rate support grant system and in the revenue-sharing formulae applied by several American states (Solomon 1988: 78). This criticism of RSC financing is mitigated, however, by two substantial *de facto* government contributions to RSC levey income:

1 Parastatal corporations will pay levies.
2 Levies are allowable as a deduction from taxable income. This means that, at the present company tax rate of 50 per cent, the government is effectively paying for half the levies.

Against this must be viewed the government's seemingly contradictory emphasis on user charges. Implicit in this is the phasing out of substantial subsidies of passenger transport (a potential RSC responsibility) as well as subsidies to local authorities in respect of such services as ambulance and fire services, health, libraries, nature reserves, and transport planning, many of which have been channelled through White local authorities to support services in African, Coloured and Indian areas in the past. Transport subsidies, made necessary by the government's own policies of locating African townships far from places of employment, are particularly critical; central government funds may eventually be channelled via RSCs, possibly in conjunction with the privatization of existing state transport services. The position regarding other subsidies remains undetermined, but the political urgency which the government attaches to the redistributional role of RSCs is such that it seems unlikely, in the last resort, to do anything to undermine it.

Centralization

The whole period 1910–83 was characterized by increased state and provincial control over local authorities (Cameron 1988: 50–1). However, recent reforms (including the creation of RSCs) have been presented as increasing devolution and local democracy. Bekker is virtually alone in identifying a key reason why the government may genuinely wish to promote devolution:

> The National Party is well aware of the enormous power wielded by a highly centralised state and is deeply concerned about the black majority assuming control of such an apparatus.
>
> (Bekker 1988: 30)

Yet other forces demand greater centralization at the present time, and there is much evidence of this in the structure of RSCs. The government's security perspective demands the securing of order and 'stability' through repression as a first priority, attention to socio-economic grievances comes second, with political reform third. This agenda leaves little room for devolution, since security measures demand strong central control (Bekker 1988: 32). As Seegers (1988: 126) succinctly puts it, 'The language of Total Onslaught is the language of centralisation.' This concern is specifically reflected in the important role now played by joint management centres (JMCs) responsible to the powerful State Security Council. JMCs are officially information-gathering bodies designed to prevent and defuse revolutionary unrest at local level. They consist of representatives of state security and other selected government departments, local authorities, and the private sector. They have become an important channel of finance for township improvement, with the priorities being decided on security grounds, as in the well-documented case of Alexandria in Johannesburg (Humphries 1988: 116).

The centralization implicit in RSCs themselves is reflected in, *inter alia*, the powers of provincial administrators, the composition of the appeal boards, and the role of the chairman as a government appointee. Although he may not vote, the latter's position as a permanent paid employee and an *ex officio* member of all committees gives him considerable power to determine the philosophy of the RSC and its implementation. As we shall see, the early operational experience of RSCs suggests that in practice the chairman's role far exceeds other centralizing features of the system.

The government regards such centralization as necessary for several reasons. First, it needs to overcome the intransigence of both liberal White authorities (in relation to its ethnic model of local government) and reactionary White authorities (in relation to the redistribution seen as a prerequisite of BLA legitimacy). The RSC chairman's role is critical in this regard. Second, the state is concerned to maintain overall control of the reform process and not let it develop a momentum of its own. The powers of the provincial administrator are important here, and so, at another level, is the intended role of JMCs in undermining alternative civic structures and simultaneously strengthening the perceived legitimacy of both BLAs and RSCs.

Legitimacy

Foremost among popular objections to RSCs are their imposition from above and their ethnic basis. The former is underlined by exclusion of community-based organizations and alternative civic structures such as area and street committees – 'embryonic organs of people's power' (Swilling 1988: 193) whose legitimacy derives from the fact that they have emerged from popular action. The suspected or known Nationalist allegiances of the chairmen also symbolize their 'top–down' character.

Whereas the government sees RSCs as a means of legitimizing BLAs, it is precisely their dependence on BLAs and the whole ethnic structure of local government which opens up RSCs to all the popular political objections levelled at BLAs themselves. In this sense RSCs reinforce what they are supposed to remedy. Such an obviously flawed conception is only possible because, as we have seen, the state has chosen to interpret hostility to BLAs in purely economic terms, whereas all major radical and moderate African, Coloured and Indian political organizations have rejected the principle of ethnically based local government.

Other legitimacy based criticisms of RSCs centre on their representation and functions. The retention by PLAs of the construction and administration of housing and township development certainly limits the redistributive potential of RSCs, but there is provision for RSC loans and/or subsidies to PLAs. Other prime concerns of township residents – unemployment, inferior education, poor health and welfare facilities – remain the concern of central government, but the upgrading of infrastructure and services are unlikely in themselves to gain legitimacy for PLAs without improvement in these basic needs. The government is not unaware of this, as a 40 per cent increase in the African education budget for 1987/8 reflects.

The weighting of voting power according to services consumed has been widely criticized as favouring Whites and undermining the redistributive aims of RScs. Such fears are heightened by the ability of White councils to ignore minority parties in choosing councillors to represent them on RSCs: Johannesburg's representatives, for example, were all Nationalists at the time of writing, despite the party's slender majority on the city council. In practice, as will be seen, White domination of RSCs has generally not precluded the intended redistribution, although this may be threatened in those RSCs which have been controlled by the extreme right-wing Conservative Party since the local elections of October 1988. However, in principle the unfairness of such a voting system will hardly lend it legitimacy, whilst the indirect nature of representation serves to distance citizens from the RSCs.

REGIONAL SERVICES COUNCILS IN PRACTICE

By mid-1989 the first eight RSCs had been functioning for two full years, and others for lesser periods. Although it is too early to judge their impact on the legitimacy of BLAs and management committees, it is possible to make a preliminary assessment of their performance. First, however, a brief consideration of their boundaries is relevant.

Boundary demarcation

Central government involvement in the demarcation process has resulted in RSCs covering much larger areas than were locally desired in most cases. The DCDP, the Minister of Finance and the 'own affairs' ministers for local government all seem to have wanted larger areas. No reasons have been given for this, but concern to get RSCs established over a wide areas as soon as possible was probably important. The provincial administrators have clearly reflected this concern, and in both the western Cape and Algoa the administrator reversed his initial decision in favour of larger RSCs – overruling the recommendation of the demarcation board in the western Cape. The Transvaal Provincial Administration originally wanted a minimum of twenty-four RSCs in the province, but settled instead for twelve. The Bloemarea RSC, which initially comprised the five local authorities of

Bloemfontein itself, was enlarged to include the whole of the southern Orange Free State in July 1989.

Although the RSC Act permits demarcation across 'homeland' borders in accordance with the government's regional development strategy, there is no instance of this having occurred as yet. This means that some 750,000 so-called 'frontier commuters' (Lemon 1982) who cross 'homeland' borders daily to work, together with their dependants and all others living in townships and squatter settlements which are functionally part of urban areas in 'White' South Africa, are excluded from RSCs. Thus the Pretoria RSC excludes well over half a million people – nearly one-third of Bophuthatswana's population – living in the Odi 1 and Moretele 1 districts, many of them squatters in the Winterveld. No effect has apparently been made by the Transvaal Provincial Administration to negotiate the inclusion of these districts, yet their exclusion deprives some of the areas most in need of upgrading from the redistributive benefits of RSCs.

Similarly, it appears that Mdantsane, a Ciskei township tributary to East London with over half a million people, together with the new Ciskeian 'capital' of Bisho and the township of Zwelitsha, both tributary to White King William's Town, will all be excluded from the Kaffraria RSC. In this case there appears to be a conflict between the RSC levies and Ciskeian tax laws, which are designed to promote industrial growth by creating a tax haven.

The sprawling semi-rural slum of Botshabelo, with a population approaching 300,000, was initially included in the Bloemarea RSC but was excised in December 1987 following its 'incorporation' into the distant and diminutive QwaQwa 'homeland'. A complex legal battle over incorporation ensued, ending with amending legislation which enabled the transfer to take place. Dr Mopedi, the Chief Minister of QwaQwa, has since repeated an earlier request in terms of the RSC Act for the RSC to provide certain services in Botshabelo on behalf of QwaQwa. An interesting point to emerge from this curious saga was the strong opposition to the inclusion of Botshabelo in the RSC from the inhabitants of Mangaung, Bloemfontein's African township, who clearly realized that their share of RSC levy income would be greatly reduced by the more pressing claims of Botshabelo.

No 'homeland' is so closely intertwined territorially and functionally with surrounding 'White' areas as KwaZulu, whose exclusion from RSCs would make a nonsense of their intended redistributive role, excluding, *inter alia*, the 1.5 million squatters of the Durban functional region. Rejection of RSCs by Chief Buthelezi and his ruling Inkatha party stemmed from the failure of the government to consult them before introducing RSCs, the loaded White vote and inadequate representation of poorer areas which need services most, the ethnic basis of local government, the minimal devolution of power involved, and the continuing lack of African participation in central government.

The deadlock was broken in March 1989, when an *ad hoc* committee of the Natal–KwaZulu Joint Executive Authority (JEA) reached agreement on a draft bill which will, when passed, make possible the provision of services on a joint, coordinated and regionalized basis in Natal and KwaZulu. Metropolitan joint services boards (MJSBs) will be established instead of RSCs. They will be responsible not to the tricameral parliament (via the provincial administrator) but to the JEA, a body which has no equivalent elsewhere in South Africa; ultimately, the Chief Minister of KwaZulu and the Administrator of Natal will be jointly responsible for resolving disputes. Unlike RSCs, MJSBs will cross 'homeland' boundaries to encompass local authorities from both Natal and KwaZulu. The bill uses the term 'poor undeveloped areas' rather than areas of 'greatest need', and the voting formula is more favourable to such areas than that in the RSC Act. However, two serious limitations must be noted: the twenty-six KwaZulu authorities to be included in the MJSBs have little authority and no budgets of their own and are thus controlled by the KwaZulu administration, whilst some 260 tribal authorities in rural KwaZulu are still excluded from the proposed MJSBs.

Service provision

Given that the provision of bulk services was the original rationale for RSCs, there is remarkably little to show for it as yet. The Central Witwatersrand RSC does intend to assume most of the permitted functions, but as a coordinator rather than a provider. Most RSCs appear to be restricting themselves to very few services for the time being and supplying these merely by entering into contracts with existing suppliers – usually White local authorities or parastatals. The Pretoria RSC has only assumed electricity supply as a regional function, primarily to enable the RSC to be constituted; the upgrading of infrastructure is clearly regarded as its real business.

In the Cape the situation is complicated by the existence of divisional councils (DCs), in effect decentralized arms of the provincial administration which subsidized them and responsible for roads, health, and other matters. Most existing and proposed boundaries of Cape RSCs follow DC boundaries, usually amalgamating two or more DCs. The two functioning Cape RSCs – western Cape and Algoa – have assumed responsibility for former DC functions and taken over their staff and equipment. This means that the RSCs are, contrary to the intention of the RSC Act, undertaking 'own affairs' functions. The same is true with respect to projects undertaken on behalf of the province such as the construction of a beach resort for Blacks at Manwabisi near Cape Town. Algoa is also providing housing, for which it is subsidized by 'own affairs' departments. Yet these same RSCs have yet to assume any regional services! Until they do, voting power is determined on the basis of services 'identified' as those to be taken over. It is therefore notionally related to services which the RSC does not provide, but unrelated to those ('own affairs') services which it does provide!

Upgrading of infrastructure

This is clearly seen as the main priority in most RSCs, and Black townships have in most cases been clearly identifiable as possessing the 'greatest need'. Decades of official policy minimizing and discouraging the development of a permanent, stable African urban population had given neither White municipalities nor, after 1971, administration boards any real incentive to provide or improve township facilities.

The Central Witwatersrand RSC, which includes Johannesburg and Soweto, made a brisk start. In July 1987 it decided to set aside R66.1m (R4 = £1), out of an estimated levy income of R70m in its first year, for infrastructure improvement in African townships. Recognizing the need to prioritize needs, it requested all towns to list their requirements. Predictably these proved widely divergent, so criteria laying down basic standards to be achieved everywhere were identified and ranked as follows:

1 Water to each property.
2 Waterborne sewage.
3 Basic all-weather roads with some stormwater control.
4 Assistance to local bodies in refuse removal.
5 Electricity: installations capable of expansion and extension to each house, but initially only street and high mast lighting.

Assessment of needs in terms of these priorities resulted in a total estimated cost of R364m – 5.2. years' expenditure at the 1987/8 rate; however, the RSC also feels that it must make some contribution to Soweto's accumulated debts of R441m, much of which is attributable to the rent boycott.

Pretoria's approach has been somewhat different. Apparently influenced by a greater White right-wing presence, it has worked out priorities according to a relative scale of needs: thus even the highly affluent district of Verwoerdburg has been allocated R8m. Even here, however, over R40m out of a total approved expenditure of R90m relates to Pretoria's two African townships, Atteridgeville and Mamelodi.

Others among the eight original RSCs were slower to identify priorities and approve budgets in their first year. East Rand did so only in February 1988, but R36m out of R42m related to African townships, chiefly for the provision of electricity, water, sewerage, roads, and stormwater drainage. The western Cape has been the slowest off the mark, and produced its 1978/8 budget only towards the end of the financial year.

The racial allocation of capital expenditure in the 1988/9 budgets of all Transvaal RSCs is summarized in Table 1.2. Nearly 70 per cent related to projects in African areas, but this percentage varied from 82.0 in Central Witwatersrand to only 6.6 in Rustenburg-Marico, where most projects are classified as 'joint'. In part this reflects variations in the racial composition of population within RSCs. The substantial size of RSC expenditures is indisputable, although in some cases a considerable proportion is carried over to the following year when projects are incomplete. More importantly for their intended legitimizing role, it must be noted that RSCs are not the only source of funding for African local authorities: to date their contribution is considerably exceeded by government and parastatal agencies, including the National Housing Fund (for Africans), the Development Bank of Southern Africa, and bridging finance provided by the provinces for African local authorities. The private sector is also making a substantial contribution, particularly through the Urban Foundation.

Voting

Whites have the required two-thirds majority of voting power on most RSCs so far established. The greatest concentrations of power

Table 1.2 Capital expenditure budgets of Transvaal regional services councils, 1988/9 (K rand)

	White	*African*	*Coloured*	*Indian*	*Joint*
East Rand	21,235 (19.9)	82,287 (77.2)	1,935 (1.8)	1,210 (1.1)	—
Central Rand	4,563 (2.5)	147,562 (82.0)	5,362 (3.0)	2,781 (1.5)	20,339 (11.1)
West Rand	5,000 (31.7)	9,400 (59.6)	950 (6.0)	425 (2.7)	—
Pretoria	30,281 (40.0)	39,755 (52.5)	3,884 (5.1)	1,831 (2.4)	—
East Vaal	3,254 (22.6)	6,699 (38.9)	36 (0.2)	107 (0.7)	5,002 (34.7)
West Vaal	3,150 (14.8)	8,170 (38.4)	331 (1.6)	—	9,642 (45.3)
Vaaldriehoek	3,809 (20.7)	14,345 (78.0)	180 (1.0)	67 (0.4)	—
Bushveld	580 (21.1)	1,931 (70.4)	100 (3.6)	132 (4.8)	—
Highveld	8,157 (48.5)	8,095 (48.1)	252 (1.5)	314 (1.9)	—
Lowveld	680 (13.6)	600 (12.0)	230 (4.6)	140 (2.8)	3,350 (67.0)
Northern Transvaal	3,325 (64.9)	1,645 (32.1)	75 (1.5)	75 (1.5)	—
Rustenburg-Marico	195 (5.6)	230 (6.6)	—	282 (8.2)	2,752 (79.6)
Total	84,815 (18.2)	320,622 (68.8)	13,286 (2.8)	7,365 (1.5)	41,086 (8.8)

Note: Figures in parentheses are percentages of the total budget of the RSC in question.

are found in Pretoria, where Pretoria (50 per cent) and Verwoerdburg (23 per cent) exceed two-thirds between them, and in the Highveld RSC where the same is true of Witbank (50 per cent) and Middleburg (26 per cent). Provincial administrators have been criticized for delimiting central business districts (CBDs) and industrial areas so as to strengthen the voting power of authorities to RSCs, such as Pretoria, and weaken that of hostile authorities such as Cape Town (33 per cent).

The largest single vote commanded by an African authority is that of the Lekoa Town Council in the Vaal Triangle (27.2 per cent), but this compares with a combined 59 per cent for the White authorities of Vereeniging and Vanderbijlpark. In Central Witwatersrand, Johannesburg commands 40.3 per cent of the voting power compared with only 17.3 per cent for the estimated 1.5 million inhabitants of Soweto. In the Northern Transvaal the 180–200,000 non-'homeland' Africans, more than two-thirds of the region's population, command a mere 0.5 per cent of the voting power because most Africans live on White farms and there is only one representative local body for Africans in the entire region (for Messina's African township, Nancefield).

White farmers have also protested about 'taxation without representation', arguing that they have little to gain from RSCs and in some cases refusing to pay levies. Some claim that they already provide their workers with the services to be offered by RSCs. Reductions in the levels of government subsidies to farmers, and the financial plight of many of them in drought-stricken areas, have fuelled their opposition. In areas like the northern Transvaal there is undoubtedly a political opposition to racial integration as well. Several RSCs have contacted farmers' associations and tried to elicit their support, with varying degrees of success. Pretoria, Rustenburg-Marico, and the Western Cape have all granted farmers 25 per cent rebates on levies, and other RSCs may follow suit.

Recognizing the problem of unrepresented rural areas, amending legislation now provides for the creation of up to three rural councils for each population group in each RSC area, with limited powers. As elected local bodies, they will be eligible for RSC representation. This curious measure not only adds additional local government bodies primarily for the purpose of electing representatives to another body, but also mean that White farmers and their African and Coloured labourers are represented on different rural councils!

In practice, voting power has so far mattered little because most

RSCs have taken few if any votes. This reflects both the power and the skill of their chairmen in getting consensus, or at least in minimizing dissent (and, of course, in getting agreed their desired programmes). In most cases it is the chairmen who have a decisive say in the organizational structure of the RSCs – whether or not there is an executive committee, for instance, and the number and role of other committees. Even the membership of committees may largely follow the suggestions of the chairman. This was the case in Algoa, for instance, where there are five committees of ten members each; all RSC members serve on a committee, and nobody on more than one – a measure which gives Africans, Coloureds, and Indians with low voting power the same representation as Whites with far greater voting power. The five committee chairmen – four Whites and one African, a businessman who chairs the Finance and Administration Committee – have delegated powers and constitute a *de facto* executive committee.

As the Conservative Party (CP) was formed to oppose the power-sharing which it believes to characterize the 1983 Constitution, it naturally opposes RSCs in principle. It has not, however, refused to serve on them, although four CP-controlled town councils in the Bloemarea RSC have sent Nationalist councillors to represent them on the RSC, apparently so as to secure the financial benefits of participation without compromising their principles! After the October 1988 local elections the CP controlled or held a veto power in ten of the twelve Transvaal RSCs, but only in four of them did it have the two-thirds majority needed to change the multi-racial character of the executive committees, which was by then the norm.

In theory CP-controlled RSCs might try to vote all or most of the funds to White areas, but appeals against such decisions by other RSC members would be likely to meet with a sympathetic response. In some rural Transvaal RSCs, the major weakness arises less from CP control than from the exclusion of most Africans because they live in 'homelands'. This allows these RSCs, most notably the northern Transvaal (see Table 1.2), to direct a considerable proportion of their income to small White communities without contravening the letter of the RSC Act. Realization of this has been a major factor encouraging CP participation.

Operational dynamics

The personal and inter-group dynamics of RSCs could be a valuable area of research, given the novelty of multi-racial decision-making bodies in South Africa. Prior to the creation of RSCs, liaison between PLAs of different race groups was virtually non-existent. Of the liaison committees which did exist, two of the more successful were in surprising places – Middleburg, in CP territory, and Morgenzon, the 'capital' of the extreme right-wing Oranje Werkerunie. Consistent with this, but equally surprising at first sight, is the emerging picture of generally positive personal relations amongst RSC members in the Transvaal and the Orange Free State, despite the dominance of Conservatives and right-wing Nationalists in White ranks. It appears that some Whites modify their attitudes when faced with actual administrative responsibilities and the acute needs of Black communities.

The role of the RSC chairmen in the Transvaal and the Orange Free State is crucial. They see themselves as more progressive than many of their RSC members, and engage in imaginative devices to ease racial and political tensions. The chairman of the Rustenburg-Marico RSC encourages close personal contact between RSC members, and even tries to involve their spouses. In Pretoria RSC meetings are held in a different town or township every month. Other RSCs have organized tours of their regions for the secretariat and officials and/or RSC members. In some instances exposure to conditions in the townships for the first time has led to a remarkable change in attitudes. It also appears that the very novelty of inter-racial contact has sharpened members' sense of participation and progress.

In the western Cape and Algoa the situation is reversed in the sense that the chairmen are more conspicuously Nationalist, party political appointees who in both cases have to deal with representatives of a city council which is more liberal, and which opposes RSCs and the segregated local authorities on which they are based. At first it seemed likely that Cape Town City Council (CTCC) would refuse to participate altogether; instead it submitted a twenty-page memorandum on how the RSC should be run, which was received as evidence of CTCC trying to take over the RSC. The association between Cape RSCs and the old DCs has given the former a conservative image. African and Coloured members in both areas are in a particularly difficult position, given the strength of popular opposition to their participation. It was in the Cape Peninsula that the lowest polls were recorded in the 1984 Coloured elections (Lemon 1985), whilst the eastern Cape has, as we have seen, experienced particularly vigorous opposition to BLAs: some Algoa townships are without councils altogether, which denies them RSC representation, although White voting power still falls below two-thirds, as in the western Cape.

Taking over the responsibilities of the DCs clearly consumed the energy of Cape RSCs in 1987 and 1988. The clear thrust of endeavour characteristic of many RSCs was lacking, especially in the western Cape RSC which took over 7,000 staff from the Cape, Stellenbosch, and Paarl DCs. Only a few senior staff were transferred to real RSC work. DC matters have tended to dominate the business of both Cape RSCs and are of little interest to the city councils; Port Elizabeth does not even receive a copy of the agendas.

There were tensions on the western Cape RSC, with some procedural wrangling and acrimony in the early stages. Some White councillors have been impatient when time is dominated by 'own affairs' matters of the old DCs or the Coloured management committees (those in the old DC areas are left with no PLA to administer their areas), which has made Coloured councillors feel ill at ease. Initially there was some hesitancy to speak out on the part of some councillors, especially Africans, but they are now participating more, especially in discussion of their own areas, where they welcome the opportunity to make their viewpoint known to Whites. All have been involved in one of the eight committees, all of which are multi-racial apart from those for liaison with White PLAs and Coloured management committees. No votes have yet been taken and decisions are presented as consensual, although this seems to mean only that objections were not voiced at the actual moment of decision.

Whilst the western Cape RSC chairman may be right in claiming the emergence of a 'positive spirit of co-operation' among members, the fundamental problem of legitimacy remains for Coloured, Indian and African members. It varies considerably in different parts of the region as three examples will show. For the representative of Kaya Mundi, the African township in Stellenbosch, it was hardly a problem; the United Democratic Front (UDF) – since subsumed into the Mass Democratic Movement) was less active than in Cape Town, and he had been elected in an exceptional 72 per cent poll in his ward. The Coloured chairman of the Macassar Management Committee was more cautiously optimistic: the UDF was not strong in his area either, and he hoped for an improved poll in the October 1988 municipal elections. The Indian representative of Rylands in Cape Town was more typical: he had been threatened on the street for his participation and feared a low poll. The position of those participating in official structures was not helped by the RSCs' late start to upgrading activities, which meant that few tangible benefits were apparent by October 1988. In the event, very low polls in the local elections reflected continuing UDF strength in the Cape Peninsula.

CONCLUSION

That RSCs are deeply flawed organizations would be difficult to contest, above all because of the ethnically segregated local government on which they are based. This is made worse by the exclusion to date of 'homeland' territory, given the functional dependence of much of the 'homeland' population on 'White' urban areas and the acute material needs of people in the settlements concerned. For the state, one suspects, this may be a lesser concern, as most such areas – apart from those in KwaZulu – have not been prominent centres of 'unrest'.

Presentation of RSCs as instruments of devolution is the opposite of the truth, although the worst fears of further centralization implicit in the RSC Act have not materialized. The servicing functions of RSCs appear largely unnecessary if not actually detrimental from the standpoint of efficiency. It seems clear that the state's real purpose is the redistributive function which RSCs are expected to perform in order to promote the legitimacy of official local government bodies. Their structure and financing is clearly intended to minimize White opposition to (and awareness of) this redistribution role.

Early experience of the first RSCs suggests that most are giving redistribution the priority intended, and spending most of their levy income on upgrading infrastructure in African, Coloured and Indian areas. Fears concerning the effects of dominant White voting power have proved largely unfounded, notwithstanding CP advances in local government. This is in part because of the key role played by appointed chairmen, but also because the multi-racial character of RSCs appears to be having a genuinely educational effect, especially in the Transvaal and the Orange Free State.

The expenditures involved will undoubtedly bring tangible benefits. The crucial question is whether these will translate into greater legitimacy for official local government structures. This may happen in areas where RSCs have worked particularly well and where extra-institutional organizations are less strong. In the eastern Cape, the Cape Peninsula, and the southern Transvaal, where alternative structures are well developed and popularly accepted, it is highly unlikely that legitimacy can be brought with such relative ease; the

1988 local elections, whilst admittedly too early a test in most areas, certainly gave no indication of this. The situation is a critical one to watch, because it will undoubtedly influence state proposals for constitutional change at regional and national levels.

NOTES

1 I should like to thank Tevia Rosmarin, Doreen Atkinson, Chris Heymanns, and Simon Bekker; the chairmen, officials, and members of the Western Cape, Algoa, and Bloemarea RSCs; Stanley Evans, Town Clerk of Cape Town, Mr J. van Wyk of the Cape Provincial Administration; Elaine Kosser, and Mauritz Norman of the South African Institute of Race Relations; Gillian Cook, Ron Davies, and Richard Fuggle of the Department of Environmental and Geographical Science, University of Cape Town; and the Human Sciences Research Council, Pretoria, for financial assistance.

2 Use of the official terms of race classification does not indicate acceptance. The term 'Black' is used to include African, Coloured and Indian, which is the usage employed by mass organizations in South Africa. Those officially designated as 'Blacks' are thus referred to here as Africans, except when using official titles such as the Black Local Authorities Act (hence the abbreviation 'BLA' in the text).

2

REFORM IN SOUTH AFRICA AND MODERNIZATION OF THE APARTHEID CITY

David Simon[1]

INTRODUCTION

The current upheavals in South Africa have generated a voluminous literature of commentary and speculation. While the various authors differ in their predictions of the speed and course of apartheid's demise, they generally agree that the events of the last seven years heralded the beginning of the end for old-style White domination. The process of change has gained new impetus since P.W. Botha's resignation as State President in late 1989. His successor, F.W. de Klerk, immediately adopted a rather different tone, pledging to incorporate Africans directly into the power structure and acknowledging that this required some form of accommodation with the hitherto banned African National Congress (ANC). Dramatic moves, such as the release of Nelson Mandela, Walter Sisulu, and fellow ANC leaders convinced many within South Africa that a watershed in the country's history had been reached.

Such developments must be examined in their broader context. Throughout the 1980s, the South African state faced growing challenges to the point where they constituted a very real threat to the regime. Two distinct elements can be discerned in this survival crisis. First, there is the immediate ('conjunctural') crisis of having to maintain day to day control in the face of mounting internal and external pressure. For example, powerful segments of the urban black[2] population mounted a sustained campaign to render the segregated townships ungovernable. The large measure of success achieved in many parts of the country prompted the state to introduce a state of emergency in mid-1986. Although brutal enforcement of emergency regulations brought a superficial calm, mass rent boycotts and other forms of defiance persisted (see Chaskalson *et al.* 1987). The second and more general element is the 'organic' crisis threatening the continued existence of the apartheid state. By the end of the 1970s, it had become clear, even to the hierarchy of the ruling National Party, that persistence of the *status quo* had become economically, politically, and socially untenable.

With the exception of parts of the White South African electorate, among whom the Conservative and Herstigte Nasionale parties and the Afrikaner Resistance Movement have led a very considerable right-wing campaign to maintain undiluted apartheid, the critical issues have become the extent, nature, and object of necessary change rather than whether change *is* required.

The Botha regime's creeping and frequently contradictory reformism in the mid- and late 1980s marked a gradual retreat from grand apartheid, a retreat now gathering pace and conviction under de Klerk. Far from this representing sudden altruism or enlightenment on the part of the government, however, it represents a survival strategy, a concerted attempt to maintain White power and influence in the face of changing circumstances. As Robinson (1986) rightly asserts, the National Party's real aim is not to remove apartheid but to modernize it. There are numerous internal and external threats to White supremacy, some largely recent (such as the mounting international sanctions campaign) and others of long standing, most importantly Black resistance. Suppression of the student-led protests in Soweto and other centres in 1976, followed in 1977 by the crackdown on the Black consciousness movement, precipitated greatly increased support for the banned ANC. Thousands of young Blacks and a few Whites fled the country, many receiving military training before returning to raise the ANC's armed struggle to new levels. Internal opposition also received new impetus with the establishment of the broadly based, nonracial United Democratic Front (UDF) in 1983. After its effective banning by the state, the UDF and allied progressive forces became known as the Mass Democratic Movement (MDM).

A subtle but very important change was also occurring in the economy over this period. Whereas apartheid had previously been instrumental in promoting capital accumulation by restricting skill acquisition largely to Whites and facilitating the exploitation of cheap Black labour, by the late 1970s this racial division of labour became a significant constraint on the increasingly sophisticated and capital-intensive economy, with its far higher skill requirements. This crucial factor underlies the significant liberalization of the labour regime, e.g. permitting Black unionization; phasing out discriminatory job reservation; allowing some provision of family accommodation for migrant labour; and changes to migrant influx control and pass laws. Increased class differentiation, primarily through the rapid growth of better skilled and property-owning Black middle classes, has characterized the last decade. Demands for political rights have increased commensurately. Juxtaposed to this since 1980 has been a worsening recession with greatly increased unemployment and poverty (Lemon 1987b; Platzky and Walker 1985). Although Africans have clearly been worst hit, no groups and classes have been immune. With Black unemployment of about 20 per cent and White working-class poverty and unemployment emerging for the first time in fifty years (B. Simon 1986), instability has grown apace. The state has thus been faced with unprecedented, though conflicting, pressures for economic, political, and social change by antagonistic racial and class interests.

Political attempts to accommodate and legitimize these changes include some concessions on bantusan citizenship laws, legalization of urban freehold tenure for Africans, and introduction in 1983 of a tricameral Constitution for Whites, Coloureds, and Asians. This excluded Africans and thus led to renewed opposition from non-collaborationists of all races. The state then sought credibility for a proposed National Council, an advisory body which would include some Africans. This was partly a response to rejection of the tricameral Constitution, and the unprecedented level of country-wide Black defiance experienced since mid-1984 (see Smith 1987), which literally made the townships ungovernable, and which has been ruthlessly suppressed by emergency powers in force since 1986.

This clearly demonstrates one element of contradiction which has characterized recent policy, namely the promulgation of stricter controls or new repressive measures in concert with more positive reforms. Such ambivalence reflects the government's dilemma, struggling on the one hand to establish credibility within the new Black middle classes without, on the other hand, being willing to alienate part of its traditional power base within the White working and middle classes. The result has generally been failure on both counts, instead achieving frustrated expectations and bitter resentment respectively. Furthermore, the state has not been able to act with unanimity. Important factional differences exist within its ranks, even at cabinet level. This is most conspicuous in rivalry between the 'state security' and military hardliners who favour direct intervention, and the 'foreign affairs' group who favour diplomacy in regional and domestic matters. More generally, too, some reactionary civil servants and even minor functionaries have on occasion acted in an obstructionist manner, failing or refusing to implement directives for change. As with any regime, the government's actions at any one time thus reflect the current balance of forces within its ranks as well as the wider political climate already noted.

As will be shown, a second problem with incremental reform is that piecemeal amendment of an all-embracing system such as that of apartheid inevitably creates conflicts and direct legal contradictions with other, unchanged elements. While this may in turn give rise to further incremental changes, a view held by advocates of reform, it is frequently not the case in practice because of severe countervailing pressures. All in all, at times the state appears to be engaged in *ad hoc* crisis management rather than pursuing a coherent strategy. In important respects, it seems to have lost the initiative, acting instead in responsive mode to extraparliamentary pressures from the MDM. The situation is exceedingly complex and in frequent flux, yet many commentators still offer simplistic analyses, predictions, or solutions.

This chapter focuses on one important facet of contemporary South Africa, namely the changes occurring in the apartheid city and the implications they hold for the future. The discussion is informed by the decolonization experience elsewhere in southern Africa and the nature of urban change which has taken place there. Whereas many existing studies have focused exclusively on urban form, the approach here examines the process of urban change in broader political and economic terms. Concern thus centres on what the changes signify in the spheres of urban production and social reproduction with particular reference to the Black underclasses. By social reproduction is meant both the short term *maintenance* and long term *renewal* of society and its constituent classes. This is integrally related to the existing mode or modes of production which, in South Africa, are dominated by private and state forms of capitalism. In the short to medium term, with which this chapter is concerned, the existence of capitalism as such is not threatened, although its precise nature and the attendant *relations* of production are more flexible. This underlies the changing class composition of urban South African society already referred to. Access to and control over the means of production are the conventional criteria used in political economy analysis. However, as will be argued, in many Third World societies including South Africa, it may be more informative to examine class relationships in terms of the key bases for securing urban social reproduction, namely land and shelter, which directly affect all urban inhabitants.

THE SOUTH AFRICAN CITY

Apartheid South Africa and its cities are often regarded as unique, defying comparison. While the extent of legal codification of racial discrimination, and the fact that internationally recognized independence was attained by a dominant White minority practising such policies are distinctive, South Africa shared a colonial history of European conquest and capitalist exploitation with much of sub-Saharan Africa. Indeed, early British 'native' policy in South Africa – including the rationale for, and implementation of, urban segregation – became the model for German and other British settlement colonies in southern, Central, and East Africa.

It was this inheritance of native reserves and urban segregation of Africans which was adopted and systematized by Afrikaner nationalists for their own purposes. The features of the distinctive apartheid city and its emergence out of the colonial segregation city in terms of the notorious 1950 and 1966 Group Area Acts, and the 1953 Reservation of Separate Amenities Act is well known (see Christopher 1983, 1987a, 1987b; Davies 1972, 1981; Davies 1971; Krige 1988; Kuper *et al.* 1958; Mabin 1986; Pirie 1984; Pirie and Hart 1985; Simon 1984; Torr 1987; Western 1981, 1985). It is important for the present analysis to note that the simplified urban structure, with its racially exclusive and unequal residential segments, educational, health and recreational facilities, was designed to minimize interracial contact, this being restricted essentially to the workplace. Even future growth was to occur outwards from each segment, thereby preserving the pattern (see Figure 2.1). This urban structure, in which White control was paramount, and where the conditions of other races mirrored their socio-political positions – and relative class status within that – both reflected and reinforced the social formation required by White domination. The appalling and overcrowded African townships established under successive Natives (Urban Areas) Acts, were designed to house at minimum cost the migrant 'temporary urban sojourners' whose cheap labour power could not be dispensed with. Restrictive conditions in terms of Section 10 of the 1945 Natives (Urban Areas) Consolidation Act, as amended in 1952, governed the granting of permanent urban residence rights. In terms of these pass laws over 100,000 arrests per annum occurred from the

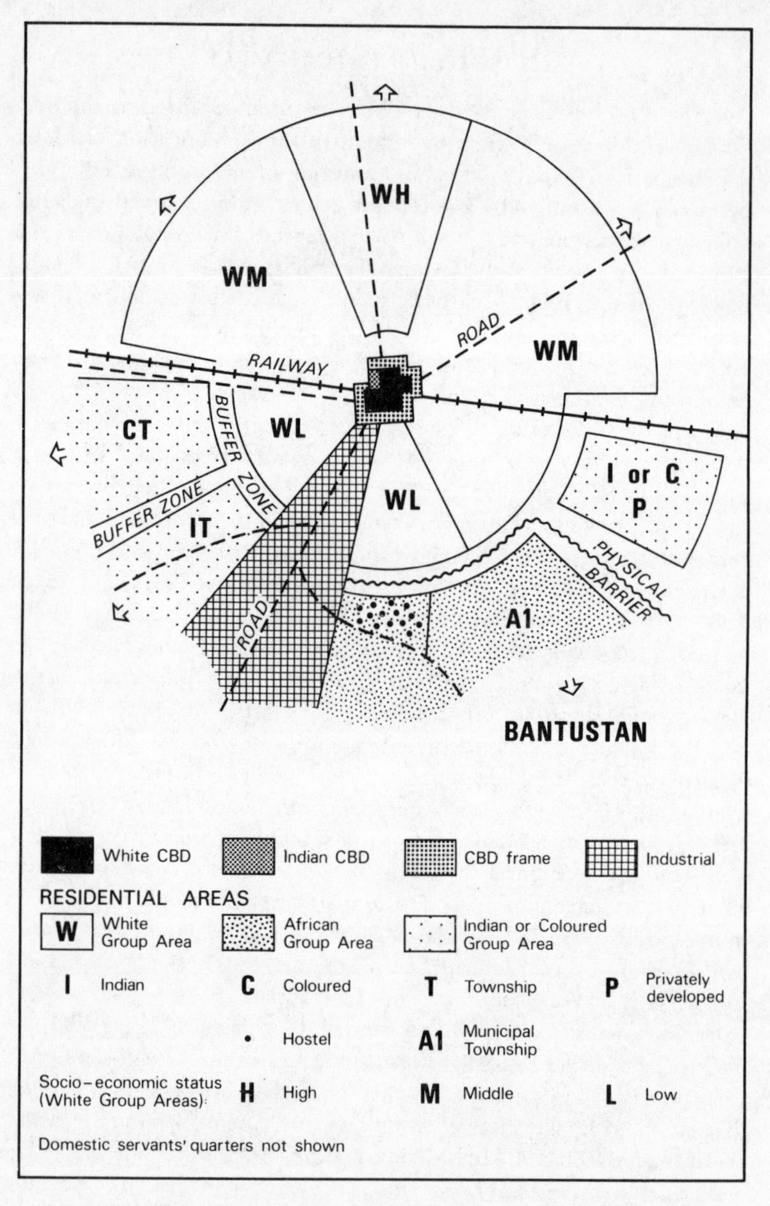

Figure 2.1 The original apartheid city model (after R.J. Davies 1981)

early 1960s until 1985, many of the offenders being forcibly sent to the bantustans (West 1982; Platzky and Walker 1985).

Over the twenty years to 1985, another 860,400 people, mainly Coloureds and Indians, and only very few Whites, were forcibly relocated within urban areas, in terms of the Group Areas Act, to create the apartheid cities and towns country-wide (Platzky and Walker 1985). While many of the removees found themselves in inferior housing in sterile, overcrowded, and distant townships, their former homes were frequently turned over to Whites at relatively low prices. In parts of Cape Town, whole areas of Chelsea cottages were renovated on this basis, creating fashionable and gentrified quarters. Other White-owned companies were involved in new township housing construction. Many Whites thus made substantial direct profits out of Group Areas removals. This process therefore represents an example of congruent interests between various segments of capital and the state in the pursuit of urban apartheid. To the above toll must be added the many thousands of Africans and Coloureds (the latter mainly in the Cape) uprooted in terms of 'slum clearance' and the 1952 Prevention of Illegal Squatting Act, and the deproclamation of townships in rural centres deemed to be within commuting distance of a bantustan. The overall spatial effect of apartheid city creation has not only been racial segregation but also increased separation between workplace and home for many working class households. With both land and capital largely controlled by Whites, the immediate burden was borne overwhelmingly by Black labour. Yet despite the huge cost in both human and monetary terms, this vast but perverse exercise in social engineering could not succeed in the face of the forces outlined earlier.

MODERNIZING THE APARTHEID CITY

The last few years have brought some significant changes affecting urbanization and urban areas. Many have not, however, so much formed elements of a sustained and integrated reform programme as *ad hoc* crisis management measures. Confusion and contradiction have abounded, reform and repression often going hand in hand. Government pronouncements and the accompanying publicity have frequently been misleading and at variance with the actual legislative changes enacted. Some of the changes may in part be responses to external pressure and the growing threat of sanctions, but the impact of such influences is difficult to determine and is unlikely to remain constant. The state has shown itself to be defiant of international opinion on many occasions in the past, especially when the US and British governments have adopted essentially benign attitudes to South Africa. Moreover, there has been sustained domestic pressure in various forms for most of the changes documented here. Foreign opinion will frequently have augmented rather than generated pressures. The contemporary situation is thus somewhat different from the sports boycott of the early 1970s, when external sanctions clearly did precipitate change.

The problem of incremental change

As had occurred earlier in Namibia (Simon 1986a), the situation has been further confused by the very complexities of apartheid's legal web. Changes to individual acts generally introduce conflicts and contradictions with others, which in turn require amendment or the granting of specific exemptions. This is evident, for example, with respect to the repeal of the Prohibition of Mixed Marriages Act and Section 16 of the Immorality Act in June 1985 (*South African Digest*, 21 June 1985). While it then became legal for people of different races to marry or live together, this contravened the Group Areas Act. Rather than amend this Act, however, the state permitted racial reclassification under the Population Registration Act: both spouses could thus be deemed of the same race and be required to live in the appropriate Group Area. Either spouse could seek reclassification although, where a White partner was involved, (s)he would almost inevitably have to take that step since dark-skinned people were still not permitted in White areas without a special permit issued by the administrator of the relevant province. In the past, such permits were generally issued only for foreign diplomats and live-in domestic servants, but a handful of exceptions have recently been made for mixed couples in the Transvaal and probably elsewhere (SAIRR 1989a). Reform has clearly not brought an end to South Africa's penchant for legal fiction. In addition to legal sanctions, social pressures and hostility from White neighbours have almost always forced mixed couples out of White Group Areas. In practice, at least in the major conurbations, the law is commonly defied, as was the case with mixed cohabitation prior to the changes. Prosecutions are still sometimes instituted, although constituting only a small fraction of all cases under the Group Areas Act (SAIRR 1989a). Proclamation of the first 'free settlement areas' in November 1989 (see p. 49) will at last enable some mixed couples to live legally without special permit. Nevertheless, the government continues to stress that it has no intention of abolishing the Population Registration Act or the Group Areas Act.

The recent changes affecting urbanization and urban areas can usefully be divided into two categories: those dealing specifically with racial zoning, and other measures. While the former have tended to grab headlines, many of the latter are far more significant in terms of their effect on urban production by, and social reproduction of, the Black majority.

Changes to racial zoning

Mixed marriages

These have already been discussed and affect relatively small numbers of people: over 130 such marriages took place in the first three months after their legalization (*South African Digest*, 11 October 1985). During 1988, thirteen Whites were reclassified as Coloureds, and fifteen Coloureds as Blacks, under the Population Registration Act. These changes are in the opposite direction to the norm, since people generally seek increased rather than decreased status, and are thus almost certainly the result of the legal and social attitudes towards mixed marriages.

Free trade areas

Sustained evidence of hardship, distortions, and inequitable trading opportunities caused by reservation of central business districts (CBDs) for exclusive White use, prompted several official investigations. In the light of their findings, Section 19 of the Group Areas Act was amended in mid-1984 to permit deproclamation of CBDs and other trading areas for commercial and professional occupation and ownership by all races. Applications can be lodged with the Group Areas Board by local authorities, individuals, groups, or the government itself. They are considered in the light of local circumstances and subject to the provisions of other relevant laws, before a recommendation is made to the government (Hudson and Sarakinsky 1986; Pirie 1986, 1987). Removals of traders under the Act had already been suspended in 1978. By early 1986 when Johannesburg, Durban, and Cape Town CBDs were officially declared the first 'free trade areas', over sixty applications had been received (*South African Digest*, 24 January and 29 February 1986). A total of ninety-seven had been proclaimed in all four provinces by the end of June 1989 (SAIRR 1989b). These are located in small country towns as well as major cities. The first free trade area to be declared outside a CBD was in Port Elizabeth. The process has not been altogether smooth, however. In such areas, the granting of freehold rights to Africans in terms of the legal changes began only in late 1987 on account of 'administrative bottlenecks'. This has impeded the ability of Africans to occupy premises since they have been unable to procure loan capital against the security of their properties (SAIRR 1987a, 1987b, 1988a).

This change – or more accurately, partial return to the pre-1950 situation – was widely welcomed by virtually all sectors of commerce and most parliamentary parties as a move towards freer enterprise and away from outmoded legislation. It was preceded, in mid-1985, by removal of the Group Areas Act restriction preventing the unsupervised employment of people in Group Areas other than their own – a provision which had been widely ignored over the previous few years. Recently, however, some opposition has been voiced, for example by Indian traders in the town of Roodepoort, west of Johannesburg. They argued against a free trade area limited to the CBD on the grounds that it represented 'an attempt to use underprivileged races to boost a dying town' because White-owned businesses were decentralizing. Apparently the whole town is, in fact, to become a free trade area (SAIRR 1988a).

After the October 1988 municipal elections, White local authorities which were controlled by the Conservative Party, most notably in the Transvaal towns of Boksburg and Carletonville, sought to oppose or reverse earlier applications for free trade area status made when the town councils were still under National Party control. However, the government has ignored such reactionary moves on the grounds that decisions were already imminent, and that deproclamation would be politically unacceptable, ruining new Black businesses and necessitating renewed evictions from Group Areas. This sharpened conflict between the government and the Conservative Party but revealed increased resolve by the former to maintain some reformist momentum (SAIRR 1989a). Moreover, the Boksburg City Council's attitude sparked a well-supported consumer boycott of White-owned shops in the city centre organized by the MDM. This ruined many shopowners but the Council resisted the combined pressure of the

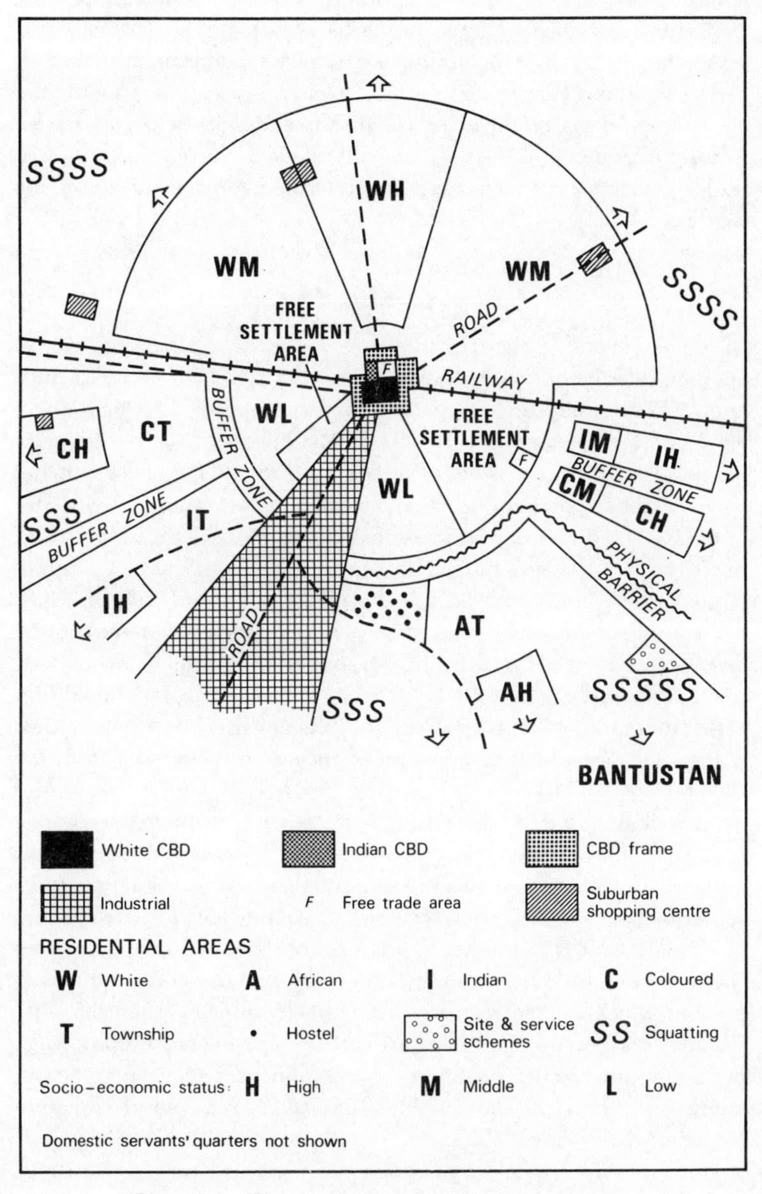

Figure 2.2 The modernized apartheid city model

MDM and affected segments of local capital. The boycott officially ended on 18 November 1989, two days after the government announced its intention of abolishing the Reservation of Separate Amenities Act (see below), in terms of which the Council had resegregated facilities including the popular public picnic area of Boksburg Lake for exclusive use of whites. Later that month, the government declared the town's CBD a free trade area (*Focus on South Africa*, December 1989). Although still relatively few in number, free trade areas merit distinction in the modernizing apartheid city model (see Figure 2.2).

Desegregation of amenities

In moves related to the free trade areas, the state has permitted the opening of drive-in cinemas and cinemas in CBDs to all races since late 1985. Initially this required ministerial approval, which is still necessary for cinemas outside CBDs. Port Elizabeth was the first city to obtain such permission in February 1986. The major cinema chains had been pressing for this concession due to falling attendances in the CBDs. Increasing white suburbanization, with the accompanying growth of decentralized business areas, and also the great popularity of home videos, are the main underlying causes. Conspicuously absent from the list of cities initially included in the cinema desegregation were Pretoria and Bloemfontein, bastions of conservative Afrikanerdom. However, following threats by international film distributors to halt supplies, Pretoria reluctantly opened its cinemas.

In 1986, several municipalities, including Cape Town, Port Elizabeth and Durban, opened their beaches, swimming pools and bus services to all races, or sought permission from the respective provincial administrators and local transportation boards to do so. A number of incidents of racial violence or insulting behaviour were subsequently reported and recurred spasmodically for some time. Obstructionism is another potential obstacle in implementing change. For example, some Johannesburg bus drivers continued to prevent Black people from boarding their vehicles after the City Council removed restrictions (SAIRR 1987a, 1987b, 1988a). Such problems also occurred after desegregation of public amenities in Namibia during the late 1970s (Simon 1986a), but declined in frequency over time.

In April 1986, remaining racial restrictions on hotels, restaurants, and accommodation establishments were removed through an amendment to the 1977 Liquor Act, which governs the sale and consumption of alcohol and which had previously been used as a key instrument in the enforcement of segregation. This move followed a decade of lobbying by the decreasingly profitable hotel industry. Previously, only specially designated 'international' hotels could serve all races.

Although by 1986 at least one government minister openly favoured abolition of the Reservation of Separate Amenities Act (*South African Digest*, 7 March 1986), this was by no means a general view. The absurdity of the situation was highlighted in January 1987 when Revd Allan Hendrickse, leader of the Coloured chamber of parliament, made a widely publicized dive into the sea from a 'Whites Only' beach in Port Elizabeth. Despite that city's application to the provincial administrator for permission to desegregate beaches a few months earlier, and the fact that the President's Council was investigating the Amenities Act and related legislation at the time, this so provoked President Botha that he humiliated Hendrickse by forcing a public apology for contravening the law, thus precipitating Hendrickse's resignation from the Cabinet. Other towns continued to enforce beach segregation with zeal, causing frequent offence and even official embarrassment. For example, the eviction of a Coloured member of the President's Council from the beach at Mossel Bay on the southern Cape coast in early 1988 precipitated a mass boycott by Black people of a major historical festival centred on the town. Organized by Allan Hendrickse's Labour Party, the boycott also received support from some liberal white quarters (SAIRR 1988a).

The government's position on the Act has been inconsistent over recent years, at one point giving an undertaking to abolish it but later retreating. In 1988, the view seemed to favour retention of segregated amenities but permitting some integration where existing segregated facilities are adequate or where particular facilities cannot be duplicated. For example, parliament requested that local authorities in Natal province open their public libraries to all races on these grounds. However, beaches continued to be the most controversial amenities, with much of the emotive debate being couched in terms of antisocial behaviour, inadequate toilet and other facilities, and fears of 'overcrowding' (SAIRR 1988a). Again, such arguments mirror the rearguard actions fought by colonial settlers elsewhere in sub-Saharan Africa and most recently in Namibia. On 16 November 1989, the new President, F.W. de Klerk, finally announced the government's commitment to abolish the entire Act, as had been recommended by the President's Council some two years earlier (see pp. 47–48). The first step was the immediate removal of racial restrictions from all public beaches. According to de Klerk, the move had been made possible because of the expenditure of R167m since 1983 on the improvement and upgrading of facilities to reduce 'the risk of friction and overcrowding'. The announcement was generally well received, and led the progressive community in Boksburg to call off its consumer boycott of White-owned undertakings.

Maintaining Group Areas

The changes already discussed, while constituting steps in the right direction, are essentially cosmetic. They benefit relatively few people, mainly the Black middle classes and White capitalist interests whose profitability had been threatened. For vast numbers of impoverished Blacks, such isolated measures do not enhance production prospects and are irrelevant to their struggles to secure a basis for social reproduction. The changes also provide little tangible 'threat' to White lifestyles or sensibilities, other than being symbolic ideological issues to the far Right. Furthermore, most of the changes to date (with the exception of the impending abolition of the Separate Amenities Act during the 1990 parliamentary session) are neither automatic nor of blanket effect, requiring state approval of individual applications (Pirie 1986, 1987). In other words, they are inconsequential in politico-economic terms. Particularly in centres like Cape Town, where segregation was less complete in the pre-apartheid era, they therefore represent only a partial return to the earlier situation rather than some major progressive initiative.

Notwithstanding pressure from diverse quarters, and the fact that Group Areas have become increasingly blurred in parts of the main cities by a combination of property market forces, a huge overall housing shortage, and freer association since abolition of the Mixed Marriages Act, Botha and de Klerk have always strongly denied any intention to abolish the Group Areas Act.[3] Routine evictions of Black people residing illegally in White Group Areas were halted in the wake of a 1982 Transvaal Supreme Court ruling, in an action brought by a support group for removees, that this was unlawful unless alternative accommodation was available. *De facto* mixing had occurred, especially in areas such as Joubert Park, Hillbrow, Berea, Yeoville, Mayfair, and Bertrams in Johannesburg, and Woodstock, Salt River, and Observatory in Cape Town, which provide relatively cheap housing close to the city centres and higher education institutions (De Coning *et al.* 1986; Rule 1988). The last few years have also seen the entry of some Black middle-class executives into the more salubrious suburbs. By far the major category of legal Black residents in White areas remains domestic servants, who are housed on employers' premises or rooftop servants' quarters above blocks of suburban flats – the so-called 'locations in the sky' (Mather 1987). In many cases this occurred with official permission, since some 90 per cent of applications for exemption from the Act's provisions during 1985 were granted.

However, despite greater apparent flexibility for a time, the state has made no secret of its desire to resume removals and maintain the Act (Pirie 1987). Accordingly, the number of police investigations under the Act increased from 1,243 in 1987 to 1,641 in 1988, with the number of prosecutions rising from 3 to 98 respectively. However, only ten of the latter resulted in convictions (SAIRR 1989a). The prosecution and conviction rates thus remain extremely low. Furthermore, the city of East London was refused permission to desegregate its White residential areas in 1986. Additionally, all new urban areas and urban property developments must still conform to Group Areas principles. Thus, although the longstanding legislative prohibition on Indians residing in the Orange Free State province was lifted in mid-1986, new Indian Group Areas are being built there, the first in Harrismith; while by mid-1987, Blue Downs, a new city exclusively for 250,000 Coloured people outside Cape Town, was under construction.

Together with segregated schooling and political representation on a 'group' (i.e. racial) basis, the Act has been cited as non-negotiable. Of necessity, therefore, the Population Registration Act, which underpins racial classification, will remain. That position has still left scope for further modifications, however. In September 1987, the long awaited President's Council report on the Group Areas, Reservation of Separate Amenities, Slums and Housing Acts was finally tabled in parliament. This recommended abolition of the Reservation of Separate Amenities Act, and amendment of the Group Areas Act to permit 'open' residential areas subject to the approval of applications by provincial administrators. Such areas should preferably be in or near CBDs, while town planning schemes should act as the main control to protect 'own' (i.e. segregated) areas. Schools in open areas would remain segregated, but tertiary education institutions could determine their own admissions policy. All land zoned for commercial, industrial, or religious purposes should be open 'in terms of the free market system'. The report was immediately rejected by both the right-wing official opposition in the White parliamentary chamber, and by the liberal White opposition and majority Coloured party, but for opposite reasons. A referendum among students on desegregating halls of residence at the elite Afrikaans university in Stellenbosch, held at around the same time, produced a small majority against the move. Even the Government's response was characteristically equivocal. Repeal of the Reservation of Separate Amenities Act was ruled out pending 'further investigation', but 'open' Group Areas were accepted in principle. A board of experts, replacing the existing Group Areas Board, would consider applications and make recommendations to the relevant minister's council and the State President, whose approval would be required. However, even such a minor change would require complex legal amendments.

In July 1988, the government published three related bills embodying its legislative plans in the light of the President's Council report. Once again, they had contradictory thrusts, exemplifying the state's unwillingness or inability to commit itself to unequivocal change. On the one hand, the Group Areas Amendment Bill sought to tighten up enforcement of urban segregation, providing for up to 25-fold increases in fines for contraventions on the part both of people selling or letting property in White areas to Blacks and of Black people living illegally in White areas. On the other hand, the Free Settlement Areas Bill would enable the declaration of unsegregated or open 'free settlement areas' on the recommendation of a board of experts as suggested by the President's Council, and subject to approval by each Cabinet of the tricameral parliament. The third bill dealt with local authority election procedures in 'free settlement areas'. 'Self-determination' remained the key, and no genuinely nonracial local councils were envisaged. What these proposed amendments amounted to, therefore, was acceptance of the reality of integrated areas where there is no realistic chance of implementing segregation, while keeping the number of such 'open' or 'free settlement' areas to a minimum and enhancing the maintenance of Group Area purity elsewhere.

Following concerted opposition both within and outside parliament, and pressure by the President's Council for major changes, the government abandoned the Group Areas Amendment Bill in February 1989. Around this time, however, several ugly incidents indicated the extent of reactionary white sentiment. Right-wing vigilantes in Johannesburg forcibly prevented an Indian family from moving into their newly purchased houses in Mayfair West, while violent scuffles broke out between the vigilantes and members of the Action Committee to Stop Evictions (ACTSTOP), who prevented the same fate befalling a family moving into Malvern. In Kraaifontein, near Cape Town, the White town council cut off water and electricity supplies to a Coloured family on the grounds that they were illegal occupants of the house, forcing them to leave. On similar grounds, the Nationalist-controlled Johannesburg City Council refused for a time to accept deposits for electricity connections to new Black applicants in White areas, but this practice was abandoned following criticism from the liberal opposition (SAIRR 1989a).

By contrast, the Free Settlement Areas Bill was passed into law in early 1989, and the first four 'free settlement areas' were declared in terms of the new Act in November 1989. These are Cape Town's District Six, Windmill Park in Boksburg, Country View at Midrand (between Johannesburg and Pretoria), and the Warwick Avenue Triangle in Durban; others are to be proclaimed in due course. There is considerable irony in the identity of at least two of the new areas. District Six was an integrated but predominantly Coloured inner city area prior to its destruction under the Slums Act and its declaration as a White Group Area, a move which spawned a sustained and almost universally supported civic boycott of redevelopment there. Windmill Park, a relatively new development in Boksburg inhabited by both White and Indian families, has never officially been declared a White Group Area. It was thus technically a 'White controlled area' under the Conservative-dominated town council which was threatening to evict the Indian residents. The impact of free settlement areas on urban social geography remains to be seen and will depend on the number, size, and relative location within urban areas of those eventually declared. However, only the relatively small number of existing 'grey' areas or other special cases are likely to attain this status in the short term. At present, the end of the apartheid city is clearly not envisaged. On the contrary, 'free settlement areas' are designed to cater for racially mixed couples and confer legality on areas where the Group Areas Act has already become unenforceable, thereby gaining legitimacy for the state in enhancing the maintenance and policing of Group Areas elsewhere. Although 'free settlement areas' have now become a feature in the modernized apartheid city model (see Figure 2.2), their impact will almost certainly be severely limited for the foreseeable future.

So long as cornerstones of apartheid such as the Population Registration, Group Areas and Reservation of Separate Amenities Acts remain on the statute book, claims by the state that apartheid is dead will rightly be ignored by all but the most gullible of critics. For it is these laws which brand people at birth, thereby determining the course and location of their lives. It is, of course, precisely because they affect everyone, and their abolition would be seen by many Whites to be surrendering White exclusivity, identity, control, and ultimately survival, that they will almost certainly remain for the foreseeable future under the present regime. The changes to date have been geared to modernizing apartheid without endangering ultimate White control. They have been consistent with the political need for incorporation of Black middle classes at the expense of impoverished 'outsiders', with a realization that the now-entrenched status of Whites is no longer dependent solely on legislative fiat in *all* spheres of life, and with the need to reduce the sheer financial cost of administering full apartheid. The state's current free market ethos and deregulatory zeal provide a convenient vehicle for achieving some of these objectives within an explicitly capitalist framework. Racial capitalism, a conception of the South African mode of production held by some radical analysts and Black consciousness groups, is therefore rapidly losing appropriateness. However, state ideology lacks legitimacy in that no consistent explanation of why these principles apply only in some cases, and why some apartheid legislation is 'morally objectionable' (i.e. rescindable) and some not, has been given.

Other changes affecting urban areas

Undoubtedly the greatest recent changes affecting South African cities arise from the state's grudging and slow acceptance of the permanence of a major urban African population. The significance of this must be understood from the perspectives of both oppressor and oppressed. For the state, the changes outlined below (see pp. 52–58) reflect the abrogation (albeit gradual and still incomplete) of a cornerstone of ideology and policy elaborated since the dawn of this century. Nevertheless, Pretoria has sought to legitimize its amendments to the *status quo* in terms of economic necessity and political astuteness. They are in keeping with the overall modernization of apartheid, and help foster the growth of an African urban bourgeoisie. For Africans, the changes provide some relief from the most hated, feared, and oppressive apartheid legislation. Harassment during their struggle for survival may consequently have been reduced, and their chances of remaining permanently in towns and cities increased. But sheer experience, suspicion, and the knowledge that much of apartheid's administrative apparatus and personnel remain in place, have tempered the response, even from within the ranks of the bourgeoisie.

In human and urban terms, the significance of these changes lies both in the sheer number of people thereby affected, and in the consequent impact on the nature and scale of future urban development. Measures affecting the stability and permanence of urban Africans fall into two categories – namely, security of access to shelter and controls on rural to urban migration. These will be discussed in turn, followed by a consideration of recent changes to the structure of urban administration.

Access to shelter

A feature of many Third World cities is that the formal or modern sector cannot provide employment for all workseekers. Industrialization is generally limited and controlled by expatriates, the state, and foreign transnational corporations. A high proportion of the urban population must therefore survive by other means. In such circumstances, access to and control over the means of social reproduction, rather than only the means of production, are crucial to survival and the processes of social differentiation and class formation (see Evers 1984; Leontidou 1985; and Simon 1989 for elaboration of these arguments). Housing is invested with use value (shelter), exchange value (as a leasable or saleable asset) and symbolic or prestige value (displaying status, values, aspirations, group affiliation, etc.). These may vary in relative significance to the occupants according to circumstance, class status, and the extent of commercialization of particular shelter submarkets. Notwithstanding some useful critiques of conventional analysis, one of the shortcomings of much neo-Marxist writing on urban residence and residential politics is a failure to recognize or distinguish all these categories (cf. Burgess 1982; Nientied and Van der Linden 1985). Situation, architecture, building materials, dwelling size, and decoration can all have symbolic importance (Martin 1974) to people of any social class quite independently of exchange value considerations, even if the process by which the symbols were created is partially or wholly capitalist in nature. Such symbols may proclaim professional, political or class status, ethnic or cultural allegiance, political affiliations and the like. This is as true in the suburbs, townships, and even squatter settlements of South Africa as elsewhere. More detailed discussion of these issues is, however, beyond the scope of this chapter.

The objective of South Africa's grand apartheid design was near-total racial segregation at all geographical scales. Concomitant with the creation of bantustans at national level was a policy of constrained African urbanization, whereby access to urban areas and urban shelter on any legal basis was (and is) tied to formal sector job availability. One of the crucial mechanisms for achieving this was state ownership of all urban African housing, a policy which necessitated abolition of pre-existing African freehold rights first granted to some permanent urban residents at the end of the last century. Coloured and Asian freehold rights were preserved, however.

The unviability of this strategy and the need to provide some security of tenure, compelled the state to introduce successively 30-, 60-, and 99-year leasehold rights for Africans during the late 1970s and early 1980s. By 1982, Africans could technically own their houses but not the land on which the houses stood (Smith 1982). Such a legal distinction had not previously existed in South African urban planning, except in the days before formalized townships when Blacks were permitted to erect their own huts or shacks in approved locations on payment of a small municipal ground rent. It is important to understand the shift in the state's position as the consequence of three interwoven factors. First, there is the state's steadily deteriorating financial position, resulting in a growing inability to underwrite reproduction of labour power through shelter provision for the burgeoning 'low income' urban population of all races but especially Coloureds and Africans. Ironically, however, the state has committed itself heavily to subsidizing reproduction of the growing middle class of public sector employees created by recent developments. Second, arising from this has been a sustained attempt to pressurize capital to increase its contribution to labour force reproduction via housing provision and payment of various levies. Finally, social differentiation among Africans and Blacks in general has been growing in response to the increased skilled labour requirements of the progressively more capital-intensive economy.

On the one hand, rising salaries fuelled demands for higher quality housing than the meagre conventional township dwellings hitherto provided by the state. On the other hand, both these economic and broader political pressures mentioned earlier, translated into state efforts to promote Black middle-class formation as a stabilizing factor in the modernizing apartheid regime. Liberal philanthropy as such was not the motive (cf. Corbett 1982; Hendler 1987; Hendler and Parnell 1987; Lea 1982; Mabin and Parnell 1983; McCarthy 1987; Wilkinson 1983) and, as with moves towards desegregation, Namibia had been used as a testbed (Simon 1986a, 1988).

African housing has therefore regained its exchange or commodity value. This both reflects and contributes to further social differentiation, and indeed class formation, for at least two reasons. First, only a minority can afford the ownership option (quite apart from some political factors which are probably further reducing the take-up rate), and second, new opportunities for capital accumulation and income earning accrue to the construction industry, small builders, and to property owners (through landlordism) in a situation of chronic shelter shortage and accelerated house building. Large White-owned supply and construction companies have long profited from the production of mass housing, an important fact frequently overlooked (see Hendler *et al.* 1986). The newness of opportunity thus applies mainly to Black-owned firms. Elite areas or even separate townships are now emerging in Black areas of towns and cities countrywide. This is especially evident with respect to the Coloured and Indian populations, where the lines between working class and mushrooming middle class are now clearly drawn not only in economic and political terms (e.g. Lemon 1987b) but increasingly also in spatial terms (see Figure 2.2).

In recognition of continuing political pressure, including a widespread rent boycott, the state agreed in 1985 to consider making provision for Africans to obtain full freehold rights in the townships. This was duly enacted by the Black Communities Development Act in mid-1986. Ownership is now permitted in urban or peri-urban townships (so-called development areas), remaining 'black spots' where pre-existing rights still apply, and in the bantustans. Citizens of the four so-called 'independent' bantustans now have the same rights of purchase as other Africans. Administrative delays in establishing property registers, arising principally out of the need to survey and demarcate individual plots, precluded actual conferral of freehold rights until the end of 1987. The process began in Soweto (Johannesburg), and by mid-February 1989 only 817 had been registered countrywide: 606 in Johannesburg, 162 in Pretoria, and 49 in Cape Town (SAIRR 1988a, 1989b). These delays were perhaps inevitable, given that townships had never been surveyed or legally subdivided while local authorities remained the sole owners. However, some bureaucratic obstructionism may also have occurred. Purchases of state housing under leasehold schemes are continuing despite the availability of freehold tenure, and altogether 107,866 houses, equivalent to 32 per cent of the available stock in African townships outside the bantustans, had been sold under the various schemes by mid-1989 (SAIRR 1988b).

Under normal circumstances, renting of houses from local authorities would remain a major form of tenure for both economic (affordability) and political reasons (opposition to buying inferior quality and frequently neglected houses in the segregated townships when many years' rent had already been paid). However, in certain areas, threats of eviction were used to accelerate purchase and some residents probably saw this as a potentially profitable step and/or a way out of the impasse created by the widespread and prolonged rent boycotts which have formed part of the largely successful campaign to make the townships ungovernable. Black councils have resorted to evictions, denial of burial plots in cemeteries, possible electricity blackouts, and the use of vigilante gangs in attempts to break the boycott (e.g. Chaskalson *et al.* 1987). However, during 1988, a new income-related formula to ensure that tenants and home purchasers did not pay more than they could afford was introduced on a voluntary basis by local authorities in the western Cape (for Coloured townships only) and Natal (for all races). Application of the formula has generally reduced rents but led to higher service charges, so the take-up rate has been slow (SAIRR 1988b, 1989b).

The granting of full freehold rights was consistent with earlier moves to foster a stable urban African bourgeoisie, and was also very likely calculated to deflate the rent boycott's momentum. Reaction varied across the political spectrum. In the case of Steve Kgame, president of the pro-government Urban Councils Association of South Africa, it was favourable: 'Ownership of property is very important. People will be assured of performance. It will be a stabilizing factor. People, knowing they own property, will not burn and destroy it.' By contrast, George Wauchope, vice-president of the radical Azanian People's Organization, was highly critical: 'The Government is shifting responsibility for control of unrest onto people it knows will maintain the status quo. The struggle is still about repossession of land. Nothing short of it will suffice' (*Guardian*, 6 December 1985).

Although this differential capital accumulation is still occurring within the confines of highly regulated state policy (Hendler 1987; Hendler *et al.* 1986; Hendler and Parnell 1987), of which the Group Areas Act is one of the most important elements in the present context, the class composition of, and basis of social reproduction in, the racially discrete segments of South African urban areas is undergoing rapid change.

Although urban land delivery and housing construction for all Black communities have accelerated markedly in recent years, the backlog continues to grow. One estimate in late 1988 suggested a shortage of 800,000 units for Blacks outside the bantustans, while officials calculate that some 200,000 houses will need to be constructed annually, at an average current cost of R20,000 each in order to meet demand (SAIRR 1988b, 1989b).

Finally, mention must be made of the growth in irregular or squatter settlements, predominantly on the periphery of large South African cities. During the early years of Nationalist rule, strenuous efforts were made under the 1952 Prevention of Illegal Squatting Act to remove all slum and squatter areas, relocating their inhabitants to formal townships or the bantustans. Although for a time apparently successful, shortages of township housing, coupled with the denial of the right to family life under the migrant labour system and deteriorating conditions in the bantustans, led to the re-emergence of significant squatting during the 1970s (see Figure 2.2). State efforts to demolish them were met with stiff resistance and adverse international publicity over celebrated cases such as Crossroads outside Cape Town and the Winterveld outside Pretoria (Lemon 1987b; Maasdorp 1982; Silk 1981; Smith 1982).

During the early 1980s, the government were forced to concede that part of Crossroads could survive and be consolidated because of the lack of alternative shelter options. This precipitated the rapid adoption of site and service schemes long after most other countries; previously these had been deemed inappropriate to the strict urban control required in apartheid cities. Even so, state-sanctioned schemes in South Africa differ from conventional World Bank-type practice in several important respects since they are still subject to other apartheid legislation and are frequently used in conjunction with forced population relocations. Secure tenure is thus obtained only at a high price (Hart and Hardie 1987). Another factor in the change of heart was undoubtedly the state's policy, previously discussed, of forcing capital and labour to assume a greater share of the costs of reproducing labour power. In other countries such motives are often not explicit (Burgess 1978). However, the South African authorities have been quite open about them. These new policies have been vigorously promoted and supported by the Urban Foundation, a non-governmental organization funded by corporate capital with the objective of helping to stabilize urban labour and defuse protest. Initial experiments were conducted at Inanda near Durban, and Kathlehong outside Carletonville in the Transvaal. The largest recent scheme is probably in Khayelitsha outside Cape Town, to which African squatters are being removed from the Greater Cape Town area (Cook 1986; Hart and Hardie 1987; Lemon 1987b).

Irregular settlements now house a growing proportion of South Africa's urban Black population, but the lack of security in the face of state coercion has certainly been influential in boosting the take-up rate of official site and service schemes, which are themselves now a distinct feature of new townships in the modernized apartheid city (see Figure 2.2). Although forced removals are now supposedly a thing of the past, the Minister of Constitutional Development and Planning admitted in September 1988 that the state still planned to remove 250,000 people living in squatter camps and townships. Another million peri-urban squatters are estimated to be under threat (*Weekly Mail*, 28 October 1988). The state assumed even greater powers than hitherto to remove squatters to new informal towns (or elsewhere) when the Prevention of Illegal Squatting Amendment Act became law in February 1989. Moreover, powers of eviction were also conferred on landowners, especially local authorities and farmers. While the central government and Transvaal provincial authorities had not yet used their powers by mid-1989, the reactionary town council in Kraaifontein near Cape Town (see p. 49) became the first to issue eviction orders under the Act (SAIRR 1989a).

Rural to urban migration

Closely related to the above changes in terms of impact on the nature and scale of urban growth, are recent modifications to the influx control regime governing rural–urban migration and urban residence rights. Apart from the sheer numbers of people affected, these changes signal the state's acknowledgement that its bantustan policy (national scale apartheid) is unsustainable.

In 1978 and 1979 two benchmark Appeal Court judgments (the *Komani* and *Rikhoto* cases) widened the basis of qualification for permanent urban residence rights of workers' dependants under

Section 10 of the Blacks (Urban Areas) Consolidation Act of 1945.[4] Furthermore, the reports of the Riekert and Wiehahn Commissions of Inquiry recommended liberalization of the labour regime, particularly stabilization of the migrant workforce, and unionization.

The state's response was to permit the labour law changes referred to in the Introduction to this chapter, while simultaneously tightening up influx controls on African urban immigration. This was effected by a three-pronged strategy. First, there has been more stringent application of the pass laws and Prevention of Illegal Squatting Act, especially by deportation to the bantustans of convicted offenders (West 1982; Giliomee and Schlemmer 1985; Platzky and Walker 1985). Second, rapid construction of large new townships has occurred, most notably Khayelitsha outside Cape Town where African housing had previously been frozen (Cook 1986; Platzky and Walker 1985). Third, Section 10 of the Urban Areas Act has been amended by the Laws on Co-operation and Development Act, 1983, so as to neutralize the effects of the *Komani* and *Rikhoto* judgments.

As one commentary remarked,

> The right to family life will hinge on the availability of approved accommodation, and the decision of officialdom to make such accommodation available. At best it places a heavy premium on the kind disposition of employers who may find it less difficult to cut through the tangle of red tape that normally confronts blacks seeking such accommodation.
>
> (*Financial Mail*, 1 July 1983: 47).

Taken together with the moves on labour law, urban tenure, and the tricameral Constitution, these measures sought to rigidify the distinctions between predominantly urban 'insiders' and predominantly rural 'outsiders' in all spheres.

In September 1985, the President's Council, an influential parliamentary advisory body, recommended abolition of influx control and its replacement with a policy of 'orderly urbanization' applicable uniformly to all races, echoing calls from organized capital and Afrikaner students among others. During the 1986 parliamentary session, the Blacks (Urban Areas) Act governing the pass laws was repealed. Orderly urbanization was henceforth to be implemented via other legislation (notably on aliens, immigration, and illegal squatting), while uniform identity documents were to be issued to all races (Lemon 1987b). In practice, influx control has been modified rather than abolished. At the same time, the Restoration of Citizenship Act was passed to enable the regaining of South African citizenship by over 8 million Africans stripped of it when the bantustans of Transkei, Bophutatswana, Venda, and Ciskei were given so-called 'independence'. They would thus hold dual nationality and regain improved employment and residence rights in South Africa. In fact, less than 25 per cent of those affected will be eligible in terms of the legal conditions; the remainder will be worse off than hitherto, since the possibility no longer exists of gaining permanent urban residence rights in terms of the repealed Blacks (Urban Areas) Act. Notwithstanding all the qualifications and limitations, these 1986 measures have somewhat relaxed influx control, thereby reversing the trend of the early 1980s and facilitating somewhat easier access to the means of urban production and social reproduction for Africans of different class and ethnic affiliations.

Urban administration

The tricameral Constitution and associated reforms have also been accompanied by a major restructuring of the state at local and regional levels. While legitimized as promoting local democracy, many of the measures implemented are in reality facilitating increased centralization and also perpetuating racist structures by creating separate bodies responsible for racially exclusive 'own' affairs and matters of common concern, the so-called 'general' affairs. Opposition on these and other grounds, both within and outside parliament, was generally overridden.

The provincial councils, hitherto elected by Whites, were abolished in mid-1986 and replaced by nominated provincial executives headed by the existing provincial administrators. People of all races are eligible for nomination. The executives are primarily responsible for administering 'general' affairs at provincial level. However, they also control African 'own' affairs, since there is no national level equivalent of the White, Coloured and Indian parliamentary chambers and ministers' councils (Lemon 1987b; Young 1988).

At the urban level, two sets of change warrant mention here. First, the discredited community councils in African townships were replaced by full town and city councils in terms of the Black Local Authorities Act of 1982. Although subject to direct control by the national Department of Co-operation and Development rather than by provincial executives, they are supposedly now on a similar footing to White, Coloured and Indian local authorities. In its own terms, this represents a significant departure from the state's previous position on political rights for Africans. Together with the modifications to influx control and the citizenship issue referred to earlier, this signifies acceptance of the permanence of large African urban populations within the new apartheid framework.

The second major element of restructuring is the introduction during 1987 of the first eight Regional Services Councils (RSCs), in terms of the Regional Services Councils Act of 1985 (see Chapter 1). These are metropolitan or region-wide bodies for the provision of 'hard' services where economies of scale are potentially important. The twenty-two possible RSC functions listed in Schedule 2 of the Act, ranging from water and electricity supply to cemeteries, abbatoirs, civil defence and tourism, are essentially local level 'general' affairs. RSCs are being managed by representatives from the racially exclusive primary local authorities (PLAs) and surviving management boards within the RSC boundaries. Membership, under the chairmanship of a state appointee, is supposedly proportional to each authority's purchase of services from the RSC, subject to a maximum of five from any one authority. Because only White municipalities possess significant central business districts and industrial areas, consumption of services in these areas is excluded from the calculations for determining RSC membership. RSCs derive their funds from three sources: sales of their services, and two business levies. The first levy (presently 0.25 per cent) is based on total remuneration paid by private and public sector employers, and the second (presently 1 per cent) derived from firms' turnovers. This financial base is consistent with one of the main stated objectives of RSCs – namely, to redistribute wealth between the various PLAs by upgrading infrastructure and services in Black areas. This also represents something of a policy change: apartheid local authority arrangements have hitherto expressly precluded cross-subsidies.

Although the state regards RSCs as an horizontal extension of local authorities, rather than a new tier of administration (see Figure 2.3), a two-tier local authority system does now exist in effect. Further details of these exceedingly complex arrangements are given by Cobbett *et al.* (1985), Lemon (1987b), Todes and Watson (1985a, 1985b, 1986a), and Young (1988) as well as in Chapter One). The process of RSC creation, boundary delimitation, and inauguration was tortuous, with the first budgets having been presented only in mid-1987. Creation of the first RSCs was in fact delayed twice, until January 1987. Much confusion, contradiction, and *ad hoc* change

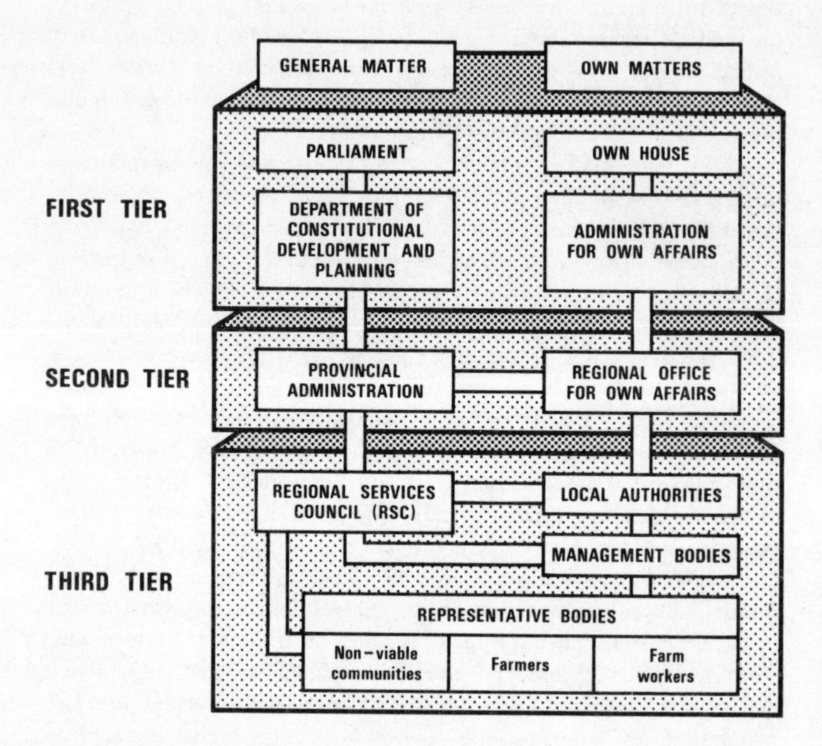

Figure 2.3 The new local and regional authority system within the overall structure of government

characterized their early operations and the entire RSC exercise has been highly controversial. Altogether sixteen councils were operational by mid-1989, although none has yet been created in Natal/KwaZulu as a result of concerted political opposition and region-specific problems. At one level, RSCs can be regarded as a logical element of reform consistent with the 1983 Constitution. The state, more particularly the Department of Constitutional Development and Planning, has invested considerable political capital in pushing the concept through on this basis and in seeking to portray the process of RSC creation as consensual. Furthermore, according to this view, RSCs bring together local authority representatives of different races on a single decision-making body, reduce the cost of local government by rationalizing service provision and exploiting economies of scale, act as a conduit for wealth redistribution, and represent decentralization of power, since some former provincial functions have been given to RSCs (see Lemon 1987b. Young 1988).

In reality, however, centralization has been increased. RSCs comprise nominated members, are chaired by state appointees, and are responsible for some former municipal functions. The state also amended legislation to permit provincial administrators to dismiss elected town and city councillors who oppose government policy, particularly the creation of RSCs.

Opposition from some quarters has been sectarian: existing White town and city councils which are losing certain functions and/or fear being outvoted by Blacks on the RSCs; and capitalist interests which are liable to the new RSC levies in addition to existing taxes. Some capitalists probably feel, however, that their longer term interests are better served by support for RSCs as a reform which might help preserve the existing (exploitative) economic system with a more acceptable face. The fundamental criticism of RSCs is certainly that their existence and role are predicated entirely on apartheid. PLAs remain racially exclusive, and votes by various White city councils to become multi-racial have been ignored by the government (SAIRR 1989b). Although spending by RSCs is now accelerating, it still remains far below their income from levies. Project expenditure budgeted by the sixteen RSCs for 1989/90 is spread between water-related infrastructure (32 per cent), roads (20 per cent), electricity (15 per cent), and various community projects (18 per cent) (SAIRR 1989b). Nevertheless, administration is proving costly, while project implementation lags behind schedule. Such problems are inevitable given RSC structures, roles and operational mechanisms, and are thus almost certain to persist.

Redistribution of wealth and integrated metropolitan development and service provision can really occur only on the basis of nonracial criteria and boundaries. It is by no means clear that major improvements in the quality of life and of conditions for social reproduction of impoverished Black communities will be possible under this system. On the contrary, the increased emphasis on wealth and supposedly free market principles as the bases of access to cities, services, and power will almost certainly increase existing disparities and tensions, resulting in greater racial and class conflict (see also Simon 1985b). RSCs have therefore been rejected by progressive organizations and even some parliamentary opposition parties and existing local authorities (Todes and Watson 1985a, 1985b, 1986a).

Since the October 1988 White municipal elections, CP-controlled local authorities have sought to frustrate the intended *modus operandi* of those RSCs on which they have control or an effective veto, by refusing to allow Black representatives to serve on the executives and by seeking to direct most RSC resources to White rather than poor Black areas. This has been possible since Black authorities do not control enough votes on RSCs to guarantee the election of Black representatives to RSC executives. Even an amendment to the Regional Services Councils Act in 1988, designed to prevent such exclusion of Blacks, is inadequate because White local authorities control the majority of seats on all RSCs (SAIRR 1989a).

The system is beset with major contradictions and practical difficulties which threaten its success. Strategically, as Todes and Watson point out:

> political reform has already played a role in focussing the attention of many community and worker organizations on the 'political'. If coloured and Indian local bodies are discredited to the extent that has occurred in the African townships then the problems of the legitimacy of the entire tricameral structure will be exacerbated.
>
> (Todes and Watson 1985b: 209)

Moreover, the undue haste, contradictions, confusion, and *ad hoc* changes which have characterized RSC implementation to date, suggest that even if the RSC concept were deemed to form part of a coherent state strategy of reform, the manner of RSC implementation more closely resembles crisis management in the search for legitimacy.

SYNTHESIS AND PROGNOSIS

The series of legislative amendments and changes analysed are of differing importance. Some benefit mainly capitalist interests and the new middle classes, or have primarily symbolic value, while others will in time affect the lives of millions. New class divisions are emerging in line with state policy to modernize apartheid, and this may actually heighten rather than lessen future conflict over the means of production and reproduction. Even taken together the changes to date have not radically altered urban form or function, although they will certainly permit more rapid urbanization of Africans.

However, it is worth pointing out that the current situation is not final: incremental change will in all likelihood continue at a variable pace dictated by circumstances. Also, the changes already enacted have set up numerous contradictions which beg further remedial action. This is inevitable when a comprehensive system like apartheid is amended piecemeal but, as elucidated earlier, the extent of change which the present regime will deem functional in its struggle for continued legitimacy is probably limited. However, the present position would seem both practically and politically untenable beyond the very short term. Increasing evidence at the time of writing (February 1990) that the government is preparing for talks with the MDM on South Africa's future (in particular the ANC and other political organizations recently unbanned) suggests that far more rapid and substantive change may be on the cards rather sooner than hitherto anticipated.

This, then, raises the inevitable question of likely future developments. I will conclude with a brief discussion, informed particularly by research in Namibia and Zimbabwe, the two most relevant countries apart from South Africa's own so-called 'independent' bantustans. Seen in comparative perspective, South Africa is currently undergoing some of the changes characteristic of decolonization. In international law, of course, South Africa has been independent since 1910 and a republic since 1961. However, most of the apparatus of segregation and apartheid which has given rise to the particularly rigid apartheid city structure was introduced and systematized after 1910 by successive White minority regimes. South Africa is unique in Africa as the only country where the White minority retained, and indeed enhanced, political power after internationally recognized independence. The Black majority have been systematically oppressed, and current events represent early stages of their assumption of at least a share of power. Their changing role within the urban political economy has been analysed earlier in this chapter. *De facto*, then, this amounts to the beginning of decolonization – not least of the minds of Whites, many of whom are becoming increasingly reactionary in their resistance to change.

Under both internal and external pressure to relinquish control over occupied Namibia, South Africa set about dismantling overt apartheid there in 1977, and by 1981 all legal racial restrictions on urban residence, property, ownership, use of public amenities, marriage or cohabitation, and influx control had been removed – albeit in a piecemeal and sometimes self-contradictory fashion. Probably because of the relative unimportance of Namibia to domestic

White South African politics, some of the most contentious issues in South Africa, such as African freehold rights, were there changed in a single step without the necessity of gradually lengthened leasehold periods en route. However, local right-wing opposition to desegregation of public amenities and urban residential areas was concerted, and Blacks and mixed couples faced intimidation or even violence on occasion, particularly in the initial two to three years. Thereafter the frequency of such incidents declined markedly (Simon 1985a, 1986, 1988). State run education and health services remained largely segregated until independence.

The urban experience of decolonization throughout Africa, most recently in Zimbabwe, Namibia and, in terms of legislative change, also the so-called 'independent' bantustans, has generally been far short of the revolutionary upheaval so feared by South African Whites. Africans have filtered into former White or Asian suburbs when housing becomes available through settler flight, emigration, and internal mobility or state expropriation. The trend is most marked in lower-middle-class areas near to African townships, commercial and industrial centres, and in certain elite areas. The remaining White and Asian minorities become spatially fairly concentrated, but only seldom have the *class* characteristics of particular suburbs changed. Symbolic urban changes (e.g. street and place names, monuments) have been rapid and ostentatious, but little structural or functional transformation of the *existing* urban fabric has generally occurred. Bricks and mortar represent substantial investment and inertia. By far the most important post-independence effects have been massive urban growth by accretion at the fringe as a result of rapid rural–urban migration, coupled with relatively slow economic growth, leading, despite often large increases in state bureaucracies, to high unemployment and underemployment and widespread urban poverty (e.g. Harvey 1987a, 1987b; Parnell 1986; Simon 1985a, 1986a, 1989; Western 1985).

The South African state has been observing developments in Namibia closely, using the country as a social laboratory. It is thus somewhat paradoxical that many of the more successful measures, such as residential desegregation, have thusfar been staunchly rejected in South Africa while notable failures such as ethnic fragmentation of health and education services have been copied with enthusiasm. This says much about the problems inherent in incremental reform and the short-term crisis management approach forced on the state in the face of White reaction on one hand and intensified

Black resistance to the *status quo* and the broad strategy of modernizing apartheid on the other. Given these conflicts and pressures, coupled with the lessons learned from beyond the Orange and Limpopo Rivers, one must conclude that South African cities will increasingly come to resemble those of other African and indeed Third World countries (cf. also Davies 1986; Lemon 1987a). Rapid urban growth and increasing class differentiation are already evident. South Africa's role as a regional or sub-imperial power within the world economy suggests that urban form and function may ultimately come to have greater similarity to other newly industrializing countries in Asia and Latin America than has hitherto been apparent, although proximity and socio-cultural history will not invalidate comparison with other southern African cities. How long this process will take, and what curious or contradictory measures will be enacted en route, will depend much on conjunctural circumstances. However, even radical revolution or transformation in South Africa may ultimately have a less dramatic impact on urban form and function, and indeed the wider political economy (Southall 1987), than is commonly assumed.

NOTES

1 Gillian Cook, Gordon Pirie, anonymous referees and participants at the Keele seminar commented helpfully on earlier drafts, while Erica Milwain drew the diagrams. Sections of this chapter have previously appeared as 'Crisis and change in South Africa: implications for the apartheid city', in *Transactions, Institute of British Geographers*, NS 14 (2): 189–205, and are reproduced here with permission of the Institute of British Geographers. This chapter was written in February 1990, immediately prior to the dramatic developments which have changed the face of South Africa. By mid-1991 most of the apartheid legislation discussed here had been repealed but a full constitutional conference still appears to be some way off. The chapter, therefore, represents a study of urban change up to the early months of F.W. de Klerk's Presidency.

2 The term 'Black' is used here to denote collectively those people classified as Coloured, Asian and Black under apartheid legislation. The legal term 'Blacks' refers to Africans.

3 A few areas, such as Lansdowne and Woodstock in Cape Town, were never fully segregated, despite state efforts. Slum clearance could not be used as a pretext as happened in District Six, while local opposition to Group Areas policy has been concerted (Garside 1987; Western 1981).

4 Surveys in unofficially mixed areas in Johannesburg found attitudes of most interviewees, irrespective of race, generally in favour of legalizing such open or 'grey' areas (De Coning *et al.* 1986; Rule 1988).

3

SOME ALTERNATIVE SCENARIOS FOR THE SOUTH AFRICAN CITY IN THE ERA OF LATE APARTHEID

Keith Beavon

INTRODUCTION

The essence of this chapter is to describe the South African metropolitan and non-metropolitan urban areas as they exist at present, and then to attempt a set of scenarios of those areas in general from 1990 to the year 2000. The situation in a particular city will then be viewed against the scenarios. Assumptions of no change, some change, or a regressive change in the socio-political system of South Africa will be used as the bases for the three scenarios. In concluding the document, attention will turn briefly to a consideration of similar scenarios for the non-metropolitan areas. Fundamental to the considerations in the chapter are what will happen in South African urban places if residential apartheid remains, is re-enforced, or is lifted partially or completely. Given the status of extra-parliamentary organizations that have a strong presence in the Black communities it follows that unless their demands are met for an ending of the state of emergency, the removal of security forces from the townships, and the release of detainees, little if any real progress will be made towards resolving or normalizing the urbanization problems in the South African meteropolitan areas.

It is important to emphasize at the outset that the form and character of the South African city has been directly moulded by segregationist and apartheid legislation for more than sixty years. The resultant South African city has deviated in several ways from the typical Western city as evidenced in North America and Australasia; most notably the poor have been removed to the urban periphery away from the inner job zone. The city of Paris, however, shares something in common with the urban geography of the apartheid city.

In the period 1852 to 1869 when 'medieval' Paris was destroyed and the 'modern' Paris was constructed the poor of that city found themselves evicted and removed from the inner city to the urban periphery during its modernization. The working class were reconcentrated in relatively high density residential districts of poor quality housing on the north-eastern side of Paris. Notwithstanding a set of particular political circumstances that prevailed in Paris in 1870 the concentration of the working class in peripheral 'ghettoes' was a significant factor in the riots, clashes, and arson that occurred during 1870 and 1871 which was known as the period of the Paris Commune. There are clearly lessons for the South African city to learn from nineteenth-century Paris but they cannot be entered into here.

Until the racial legislation referred to above is repealed – most notably the Population Registration Act, the Group Areas Act, as well as the Slums Act and the Squatters Act as presently formulated – there is little chance that the South African city will provide an acceptable and desirable urban form for all, or at least the majority, of its citizens. Indeed so entrenched are the effects of the aforementioned areas of legislation that even if they were repealed overnight the resultant changes in the form, structure, and character of the urban places would take some considerable time. Meanwhile the South African urban areas are menaced by the four horsemen of the urban apocalypse: namely, poverty, overcrowding, homelessness, and unemployment – all of which impact most directly upon the urban Black population.

This chapter assumes that its readers are reasonably well informed on South Africa. In the main only the extremes of the racial spectrum are considered – namely, the Black population and the Whites. It is assumed that if the 'urban problems' can be resolved for those two groups then there will be a natural accommodation of the presently constituted Indian and Coloured population groups. The significance of any ideas in the chapter must necessarily outweigh the provision of a detailed database. In general, the data used are those which are commonly accepted as being representative of the present situation in particular areas and places.

THE METROPOLITAN URBAN AREAS

The South African urban system can be divided into three main categories: the major metropolitan areas and the non-metropolitan areas, with a small intermediate group of minor metropolitan areas. For the purposes of this chapter the term 'metropolitan area' will apply henceforth to the major metropolitan areas only. The four major metropolitan areas of South Africa are Pretoria–Witwatersrand–Vereeniging (PWV), Cape Town and the Peninsula (CTP), Durban–Pinetown–Pietermaritzburg (DPP), and Port Elizabeth–Uitenhage (PEU). The dominance of the major metropolitan areas in the South African economy is reflected not only in the fact that they contribute 70 per cent of the Gross Geographic Product (GGP) of the Republic, with the PWV area alone contributing almost 43 per cent of the GGP, but it is in these four areas that most of the job opportunities of the country are concentrated. The four major metropolitan areas contain 46.7 per cent of the total South African population of 20.5 million. If the so-called 'homelands' are included then the major metropolitan areas still reflect 32.6 per cent of the 29.5 million people within the 'old' South African boundaries. More significant, however, 71 per cent of the urban Black population outside of the 'homelands' is resident within the boundaries of the four metropolitan areas. It is estimated that by the year 2000 there will be 25 million Black people living in the urban areas.

Common factors

There are a number of common factors associated with the four metropolitan areas that are relevant to the scenarios that will be presented in this chapter. They all represent areas of agglomeration that have emerged from outward growth around individual urban nuclei. Each of the four areas contains at least one urban place which at one time or another had Black, Coloured, or Indian freehold rights in townships or suburbs and which have subsequently been declared White group areas: Sophiatown to Triomf in Johannesburg, District Six to a vacant area in Cape Town, and South End in Port Elizabeth. Therefore all the metropolitan areas contain groups of people who have been relocated from areas where they were born, or had chosen to live, to areas which had been assigned to them. All the urban places in the four metropolitan areas have peripheral Black locations/townships, created in response to the 1923 Natives (Urban Areas) Act and which, at least in part, abut newly emerged White suburbs/townships of other urban places within their metropolitan area. All the metropolitan areas have a housing shortage for their Black/Coloured/Indian populations as exhibited by the existence of substantial communities of squatters within or adjacent to the metropolitan boundaries.

Of particular significance is the composition of population by race and age. The White population in the metropolitan areas is essentially an ageing population whereas in the Black population almost 40 per cent are below the age of 18. Implicit in the population statistic is an increasingly more conservative White population and an increasingly frustrated group of young Black South Africans. All the core urban places in each of the metropolitan areas have both inner and outer industrial zones that in turn necessitates journeys to work which are of both short and long duration for the Black workforce now resident on the peripheries of the individual municipalities within the metropolitan areas. Notwithstanding their relative 'youth', the core areas of each of the major urban places in the metropolitan areas are now old enough to contain inner zones of residential blight that are ripe for renewal. In almost all cases the old inner residential zones are characterized by low population densities.

Although the above description of common factors in the metropolitan areas implies a uniformity of perception for all the residents, such is not the case. Given the relative deprivation of Black people and the strict controls that until very recently precluded them from using many facilities in the metropolises in naturally follows that even now their perceptions of the advantages and disadvantages of living in a metropolitan area must differ from those of White people. Even amongst the White population there will be different perceptions of life in the metropolitan areas which in turn will have a bearing on the future scenarios that are presented. Before considering a set of scenarios based upon the perceptions and consequent expectations of people in the metropolitan areas, attention first turns to a more general consideration of the inherent problems and opportunities which are a feature of South African metropolises.

PROBLEMS, ADVANTAGES, AND OPPORTUNITIES IN THE METROPOLITAN AREAS

Put in the simplest possible terms, the problems confronting the South African metropolitan areas are poverty, shortages and high costs of low income houses, unemployment, racial restrictions, and the denial of meaningful political participation to the majority of urban residents. This sorry mix of trouble is further complicated by the desires of government to create satellite Black towns or city states which will merely exacerbate the frustrations and problems that already bedevil the metropolitan areas. For example, although the

Table 3.1 The only permissible trades, businesses, and professions in Black townships between 1923 and 1977

Type	*Permitted from*	*Type*	*Permitted from*	*Type*	*Permitted from*
General dealer	1923	Medical practitioner	1976	Dealer in scrap bottles, bones and used goods	1977
Eating-house	1923	Attorney	1976	Dealer in electical equipment (not repairs)	1977
Milkshop	1923	Barber or hairdresser	1976	Dealer in hardware and building equipment	1977
Restaurant	1923	Cobbler	1976	Commercial Rep.	1977
Butcher	1923	Street photographer	1976	Homeopath	1977
Hawker	1923	Fishmonger and fish frier	1976	Wood and coal	1977
Fruit, vegetable and plant dealer	1923	Laundry or dry-cleaning receiving depot	1976	Watchmaker	1977
Chemist and druggist	1976	Launderer or dry-cleaner	1976	Tinker	1977
Funeral undertaker	1976	Optometrist	1977	Tailor/dressmaker/outfitter	1977
Cycle dealer	1976	Auditor	1977	Crèche or nursery school	1977
Dealer in household, patent and proprietary medicines	1976	Warehouse	1977	Coffee bar	1977
Dealer in aerated or mineral water	1976	Pawnbroker	1977	Herbalist	1977
Kennel or pets boarding establishment or salon	1976	Radio dealer (not repairs)	1977	Boarding-house	1977
Livery stable or riding school	1976	Accountant	1977	Furniture	1977
Cafe keeper	1976	Stationer	1977	Miller	1977
Physical culture, health, or beauty centre	1976	Debt collector and tracer	1977	Motor driving school	1977
Filling station	1976	Insurance Rep.	1977	Fencing	1977
Passenger transport undertaking	1976	Tinsmith	1977	Disinfector or fumigator	1977
		Bookseller	1977	Sport shop	1977
		Optician	1977	Upholsterer	1977
		Auctioneer	1977	Petshop	1977
		Caterer	1977	Transport (goods)	1977
		Dentist	1977		
		Place of entertainment	1977		
		Chiropractor	1977		

Sources: *Government Gazette*, Notices R1036, June 1968; R764, May 1976; R2292, November 1977.

Note: Restrictions on Black businesses were lifted in 1977.

central business districts (CBDs) of many metropolitan areas have now opened to Black entrepreneurs allowing them an opportunity to develop businesses in a free market system (see Table 3.1), the government now wishes to create 'self-contained' Black cities. The whole is aggravated by the fact that overall the levels of education for the population at large are low and the content of the education does little to inform them of the history and nature of the real urban problems facing South Africa. In order to provide a reasonable foundation for the requested scenarios it is necessary to discuss the problems and advantages of the metropolitan areas from a neutral position. Thereafter, attention will be given to the varying and more partisan perceptions of life and opportunity in the metropolitan areas which probably inform different population groups.

General advantages, disadvantages, and urban problems

There are naturally both advantages and disadvantages associated with living in the metropolitan areas of South Africa. Broadly speaking it can be argued that the White zones of metropolitan areas in South Africa approximate closely in character style and opportunity to the metropolitan areas of First World countries. Given the concentration of large populations in relatively small areas there are many economies of scale that facilitate and support good health, recreational, educational, commercial, and industrial concerns and institutions in the South African metropolitan areas. All metropolitan residents can therefore anticipate a wide variety of shopping centres providing a massive array of consumer goods at competitive prices, good theatres and cinemas, a variety of pre-school, primary, and secondary education institutions both private and state controlled, as well as both private and state medical practices and facilities. In addition the metropolitan areas all boast at least one university together with other tertiary education centres including colleges of education and technikons.

Because of their age, all the metropolitan areas contain blighted residential zones that are ripe for residential renewal. The possibility exists, however, of the old areas being turned to advantage. Renewing blighted, low-density areas can bring about acceptable higher density, inner-city low-income residential areas which need not be restricted to Whites alone. Given the nature of the South African city, with both inner and outer industrial zones, the development of inner-city working-class areas would go a long way towards alleviating many of the tedious journeys to work for Black workers.

There are clearly a wide variety of urban problems associated with metropolitan areas. There is little need to set out the detail of the current urban problems but rather to list in brief form those problems that are considered germane to the discussion set out in this chapter.

1 In all the metropolitan areas there are high and rising crime rates that are more closely associated with increasing unemployment rather than with racial mixing in the inner areas.
2 The present zoning on racial lines allows insufficient land for low-income housing for Blacks.
3 Lack of urban land for Blacks is aggravated by the low-rise and low-density residential sprawl of White suburbs.
4 There is an inadequate stock of housing for the low and lower-middle-class Black population in particular. The shortage nation-wide is estimated to be between 400,000 and 800,000 units.
5 Many would argue that the single major problem in the metropolitan areas is how to successfully elimate the factors that give rise to large peri-urban squatter settlements.
6 Frustrations have been created by the Group Areas Act acting against an improvement in the job opportunities for Black people, their rising aspirations and improved levels of education.
7 Increasing levels of air pollution are occurring particularly from smoke emissions. Currently there are 260,000 homes in Soweto alone that are without electricity and which must still make use of coal fires for heating and cooking.
8 The continuing low-density sprawl increases in the cost of the journey to work.
9 The sprawl also increases the perceived and real demand for decentralized business centres which in turn places the central business district under threat of deterioration.
10 Deteriorating social conditions for the lower income groups and the increasing 'army' of the unemployed. 'Soup kitchens', which have even become a feature of daily life in some of the poverty stricken White suburbs in metropolitan areas, have long been found in the Black townships.
11 In the South African metropolis there are many instances of White residential sprawl around areas that were once seen as the Black periphery. Given racial zoning of land, coupled with differential growth rates in White and Black areas, urban sprawl has had the effect of creating zones of convergence between high and

low-income areas occupied by different racial groups. Given the tensions that appear to arise most readily between low-income White residents and Black residents in metropolitan areas the potential for conflict between such adjacent groups increases. Similar conflict, although not necessarily physical conflict, can arise where low-income Black residential areas begin to converge on high-income White/Coloured/Indian areas.

Unequal opportunities in the metropolitan areas

Not only are metropolitan areas characterized by a wide variety of facilities, such as those just outlined above, but the conglomeration of large populations also provides numerous opportunities for eager entrepreneurs who wish to carve out their own income-earning niche or empire. In a society of equal opportunity and good government there would be considerably more advantages than disadvantages associated with metropolitan areas. Such, however, is not the case in South Africa where segregation and apartheid have given rise to many serious contradictions that alter the perceived and actual advantages for different groups. Up until comparatively recently the following contradictions were prevalent:

1 Whites enjoyed a wide variety of job opportunities; Black people were restricted to certain job types.
2 White entrepreneurs were entitled to operate more or less at will in a free-market economy, whereas Blacks were severely restricted in this respect until as late as December 1977 (see Table 3.1).
3 Whites enjoyed a wide variety of housing types and locations; Blacks were restricted to a poor selection of housing types, mostly of the 'matchbox' variety in Black locations or townships on the periphery of the urban areas.
4 Whites enjoyed a choice of schools and schooling for their children, most of it at nominal or no cost at all; Blacks were restricted to neighbourhood schools that were controlled by the Department of Bantu Education (now the Department of Education and Training) for which books and stationery had to be purchased and school fees had to be paid. In some instances Black children had access to expensive and frequently distant, private schools.
5 Whites enjoyed access on academic merit to universities and other well-endowed tertiary education facilities; Blacks were either not permitted to attend or could attend only with the permission of a minister.

6 The wide variety of cultural and recreational facilities enjoyed by the White metropolitans were for a long time denied to their fellow Black metropolitans.

The lifting of many of the restrictions implied above, the introduction of home-ownership schemes for the Black population, and the small improvements in the availability of tertiary education have all assisted in raising the standard of living for many Black people. In so doing their anticipations of a better life in the cities has also increased, thereby contributing to the growing pressure for concomitant change in other facets of the urban milieu.

A major advantage in the present metropolitan areas in South Africa is that they reflect a cosmopolitan character and this in itself could be gauged as supportive for any attempts to declare the metropolitan areas open to all citizens. Evidence supporting such a view already exists in parts of all the metropolitan areas of South Africa. Certainly in Cape Town, Durban, and Johannesburg – core cities in three of the metropolitan areas – Black/Coloured/Indian people have moved in significant numbers into the inner high-rise residential zones of those cities. Estimates of the numbers involved vary but appear significant, to wit 30,000 to 50,000 for Hillbrow and its environs, where some 2,500 housing units are owned by 'Black' people and the rest rented; and as many as 25,000 Indians in Mayfair, Johannesburg.

The opening of the CBDs to other than White entrepreneurs has also been an important and positive step towards a more normalized city. Unfortunately, as revealed in recent elections, all four metropolitan areas are dominated by right- or far right-wing White groups with only a few small enclaves of progressive-thinking Whites. Consequently it remains a moot point as to whether developments in the future will build upon what some regard as the positive movements that have taken place or whether the same movements will be stopped because they are regarded as negative.

One of the major spatial disadvantages of the apartheid city is the fact that much of the real poverty, and the poor urban environments associated with Black/Coloured/Indian residential areas, is literally hidden from the view of most Whites. As such the 'inhumanity' of the elite group, who are responsible for instituting and/or maintaining the legislation that sustains the urban apartheid, is not sufficiently visible on a day-to-day basis. Consequently, many White people believe there are few if any urban apartheid problems.

Factors affecting the dynamics of growth

There are several factors that must be considered in attempting to assess what the metropolitan areas might look like in the future. The restriction of land ownership and occupancy under the Group Areas Act meant that for a large section of the population there was increasing population pressure. The irony is that because of limited reforms to date the pressures are now increasing. The opening up of the job market to people of all races has meant that a significant number of Black people have moved into reasonably paid, or even well paid, jobs. With a rise in personal income has come a desire for the better things in life. Not least is the desire of the aspirant Black bourgeois and petit bourgeoisie to live closer to their places of work, and to live in more pleasant surroundings. In recent years there has also been a desire to flee the turmoil that has been centred in and around the townships.

An important factor that effects the dynamics of the South African metropolis is the housing subsidy for first time buyers. The subsidy is specifically for the purchase of *new houses*. Apart from being of assistance to the building trade, the subsidy provides an incentive for White people to move out of the cheaper inner city areas where they rented houses or apartments, and to purchase units in the new and relatively cheap housing estates on the metropolitan periphery. The move of Whites to the periphery left vacancies in the inner cheaper residential areas that could be filled by Blacks if it had not been for the Group Areas Act. In some suburbs the Black/Indian/Coloured population has indeed taken up some of the vacant space.

URBAN APARTHEID PERPETUATED: THREE SCENARIOS

Attention now turns to a consideration of what might occur in the metropolitan areas if the present system of urban apartheid were to persist through to 2000. In attempting the task it is necessary to take note of three sets of circumstances. First, in the pre-election period urban apartheid, as perceived by many, appeared to be weakening and there was widespread anticipation that the Group Areas Act was on the verge of being scrapped. Second, the noticeable swing to the right of the political spectrum by White South Africans as evidenced in the recent election results, has sustained a markedly more pessimistic view. As a result not only did many people now believe that the Group Areas Act would not be scrapped but they believed that it might be more rigorously re-enforced. Finally, there are others who believe that the state has recognized the pressure both from within its own ranks and from international moderates, in its move to scrap the Group Areas Act which is perceived to be the cornerstone of all urban apartheid.

The government has repeatedly stated that it is proceeding and will continue to proceed with a reform programme and recognizes that apartheid must go. The government is however still fearful of the right- and far-right-wing political groups. Given its position between the Scylla of mounting expectations both locally and abroad, and the Charybdis of the right- and far-right-wing fears three options emerge for the growing metropolitan areas under apartheid, notwithstanding recent changes.

Two of the scenarios are based on the assumption that the government will adopt policies that are straightforward and that they will live with the reactions that are generated. The third is based on the assumption that the government will adopt a seemingly honourable policy but will administer it with a sleight of hand which will allow it to cock-a-snook at outside criticisms but at the same time continue to maintain urban apartheid. As a consequence there are three possible apartheid scenarios that need to be sketched.

Scenario A: the blind-eye approach to grey areas

The first scenario is based on the assumption that the government will adopt a 'blind-eye' approach to the existence and continued existence of the grey areas. It is also assumed that the government will allow the present grey areas to become a darker shade of grey. The government will be the sole arbiter of how dark the grey areas will be allowed to become and will at some stage initiate restrictions that will prevent the dark grey areas from becoming Black areas. Within the scenario as a whole a number of issues, actions, and reactions emerge and the scenario can now be considered in more detail.

Underlying the first scenario is the fact that the emergence of the grey areas over the past several years was perceived by many as the stepping stone which could be used in the lifting of the Group Areas Act. Unfortunately there is now little doubt that the existence of grey areas has been partially instrumental in once again raising the spectre of *swart gevaar* amongst White people, more particularly amongst women and the older White people. In Scenario A the underlying dynamics are a continuing movement of Black people into the existing grey areas without any new grey areas being opened up beyond the inner residential zones. With the economy limping along and taking intermittent body-blows brought about by strikes, disinvestment, and a weak rand, unemployment will continue to increase both in absolute and in relative terms as population increases. As unemployment increases, petty and more serious crime will also increase. There is a distinct posibility that in the minds of a significant number of White people the increase in the crime rate will be perceived to link directly to the increase in inner city Black population.

As the movement of Black people, in Scenario A, will have taken place under a blind-eye approach to the Group Areas Act, Black occupation of White residential accommodation will in fact be illegal. The implications of being an illegal Black resident in a White area are serious. As an illegal resident you can easily be exploited by a variety of ruthless persons not the least of whom might be your landlord: the 'bottom line' being exposure and removal. For example, as an illegal you will not be able to officially have a telephone in your name, you will be hesitant of divulging your address to any government or state agency including the post office, and the television and motor licensing authorities. As such you will expose yourself to further potential problems with the state/local authorities. Parents with a young family would certainly have problems finding a school that is both suitable and proximate. As a consequence, experience of the 'good life' will be frustrated and it would be natural for this to take the form of smouldering resentment and anger towards the authorities in general and the apartheid system in particular.

The possible exploitation of Black tenants by White landlords deserves a little more attention. Not only might some landlords of multiple dwelling units resort to rack-renting practices but estate agents may well refuse to place additional residential units on their books with the result that the number of occupants per unit could increase. White tenants unable to sell their properties to Blacks but wanting to move out of areas where Black occupation is increasing will resort to privately renting the same property to Blacks. The Whites will then flee the inner residential areas for the comparative safety of the more expensive and more distant middle and outer residential zones. Without an increase in the amount of land for Black housing in Black areas, without substantial additions of housing units, and without considerable financial assistance to the lowest income renters and the middle income home purchasers, the pressures of overcrowding, inadequate housing, and unemployment will continue to fuel the voluntary and involuntary resistance to the apartheid regime. Consequently, unrest in the townships will increase. Government intervention in the form of police and troops will at best only contain such unrest from time to time. Where Black residential areas impinge on White areas cross-boundary 'raids' can be expected, particularly in the form of White vigilante groups or by White sponsored Black vigilantes. (It is widely believed that vigilante action of both the types mentioned has already taken place, examples being in Munsieville and Crossroads.)

A point must eventually be reached in Scenario A where attempts to provide, maintain, and charge out the costs of basic services such as refuse removal and water and electricity supplies, and to provide health care and education, will simply break down in the Black townships. The view has been expressed in a recent Syncom report that population growth and urbanization have already created an explosion in the demand for services that is beyond the scale of what the country can afford. An escalation of violent reaction to a dcteriorating social system is a distinct possibility and, whether justifiable or not, could then begin to spill over from the Black townships into the grey areas and the interface areas. If the Black pressure groups were to become effective within the (by then) dark-grey areas, those areas could also become 'no-go' areas for Whites from time to time, just as is now the case in the Black townships. Under such circumstances the security agencies of the State are sure to intervene. Although the possibility of a Beirut/Belfast situation may appear to loom large the resultant situation will more likely resemble that at Seoul in 1987: popular demonstrations and violent confrontation with the 'law' enforcement agencies in the inner city districts. Before such an 'end game' is played out in Scenario A however, a 'fail-safe' gambit by the state would most probably be introduced. Once the shade of the grey areas had reached a determined or predetermined level the state would simply commence enforcing the provisions of the Group Areas Act. By insisting on occupation by permit only, and by then withholding additional permits, the 'shade' of grey could be held at a particular level. The ramifications of such an act would once again be to increase the living pressures in the Black/Coloured/Indian areas with their burgeoning populations of predominantly young people.

Scenario B: an re-enforcement of the Group Areas Act provisions

The alternative to the continued existence of illegal grey areas associated with the blind-eye attitude of the state would be the vigorous enforcement of urban apartheid by its various administrative and security arms. There have been ominous signs that such could be the case: evidenced by the reported serving of eviction notices on Black and Indian people living in grey areas of Durban (Albert Park area) and Johannesburg (Hillbrow and the city centre). If Black people are removed from the grey areas they will have to find accommodation elsewhere. Certainly, if they are removed to existing Black/ Coloured/Indian townships their readmission into those areas can only add to the existing problems of overcrowding and shortage of accommodation.

From the point of view of those Whites in the inner areas who feared the Black 'invasion' and who perceived the Black occupation as contributing to increased crime rates, the removal of Blacks will be welcomed. Once again the results of the recent elections would tend to suggest that such would be the attitude of a very large percentage of White people in any residential area. If, in terms of Scenario B, there are a significant number of removals of illegal Black residents from the White areas, it will be regarded by many Whites as a desirable display of force (*kragdadigheid*). The net result after a period of turmoil will then be one of outward calm in the relieved White areas. The calm, however, would be deceptive for it would not be reciprocated in the crowded and overcrowded ghetto areas of the Black townships where no relief from accommodation pressure would have been forthcoming. Certainly mumblings about self-contained autonomy and city-states, and promises of development funds to be collected by the as yet untried and unproved regional service councils (RSCs), will do little if anything to alleviate the growing sense of bitterness and resentment amongst the subjected majority towards the White elite. Therefore, any calm that the re-enforced urban apartheid might apparently bring about would be deceptive. Such a calm could be equated with the superficial calm experienced by the Aryan inhabitants of occupied Budapest in 1944 when around them, but out of sight of the cities, the Nazi administrators were busy implementing their form of a 'final solution'. In the South African case the 'solution' would be taking the form of increased population pressure in existing townships and/or the creation of similar 'ghettos' elsewhere.

A third aspect of Scenario B also needs to be considered: namely, the conditions of life styles in the Black/Coloured/Indian townships if urban apartheid continues to be practised as the population grows and the economy stutters along. Given the tight boundaries that already enclose such areas there is no real room for outward expansion. The only relief for Black residential demands that could be brought about within a re-enforced apartheid paradigm would be to develop new Black urban places/ghettos in the interstitial space between urban places of the metropolitan areas. The first suggestion along such lines was included in the so-called PWV Guideline Plan and took the form of a township instantly dubbed Norweto. So great was the outcry generated against the proposal that it has for the time being been dropped or shelved. Yet in a country where state 'machinery' has operated for years with scant regard to public opinion it certainly is possible that the Norweto idea could yet become a reality. Notwithstanding the varied reasons for White objections to Norweto the bottom line would appear to be White rejection of a Black township in an areas that is perceived to be a White preserve.

Scenario C: repealing the Group Areas Act but maintaining apartheid

As indicated in the preamble to the discussion the third of the scenarios is based on the assumption that the government might attempt some political shift in order to appease outside criticism while at the same time maintaining White group areas. Repeal of the Group Areas Act alone will not bring about an end to urban apartheid. Suppose the Group Areas Act was repealed, then in theory it would be possible for Black people legally to purchase and rent property in the current White space of the metropolitan areas. As already predicted by members of the Institute of Estate Agents, repeal of the Act could see a rise in the price of White-owned properties. In Mayfair the market value of White-owned properties sold to Indians sometimes increased by two or three times in price. A price hike would reflect either the anticipation of capital gains through selling the property to Black/Coloured/Indian people or could simply reflect a defence of White property by putting it beyond the reach of all but the very wealthiest Black/Coloured/Indian people. Price hikes notwithstanding, it is more than likely that the inner residential areas would be increasingly occupied by Black people for reasons already discussed. Certainly price hikes associated with the spread of grey areas have already occurred in some parts of Johannesburg.

Even after the Group Areas Act is repealed, the Government and its agencies and allies (the local authorities and some private landowners) will still have access to two other legislative instruments which if manipulated could effectively serve the ends of those who wished to retain the White areas for White people. The two instruments are the Slums Act and the Squatters Act. Just as provisions in the Squatters Act can be used to replace the influx control system so can the Slums Act be used to replace the Group Areas Act in many instances.

If Scenario C were the strategy to be selected by the government then the initial developments could follow a trajectory similar to that set out for Scenario A. It follows that some 'darkening' of inner city grey areas could well be permitted before any changes were considered necessary. If in fact the prices and rents of inner city residential units and properties did rise after repeal of the Group Areas Act, then a counter strategy from those legally moving into the grey areas could be to share units. As such, the population densities could well give rise to conditions that would make implementation of the provisions of the Slums Act appropriate. Furthermore, if unscrupulous landlords were to exploit the financial circumstances of many Black residents it could take the form of rack-renting coupled with low maintenance on the buildings. Once again circumstances could arise which would be cause for implementing the provisions of the Slums Act.

The Slums Act of 1934 has been amended many times. As presently formulated, a local authority can declare an area, a neighbourhood, or even a specified property to be a slum. Slum conditions can be determined to exist when premises *inter alia* are considered dirty, dangerous to health, unsafe, or overcrowded. Once declared a slum, the premises can be demolished or expropriated, or the tenants/owners can be evicted. If the owner fails to respond then the local authority can remove the nuisance. Whereas the Slums Act contains honourable intentions it could, if necessary, be used to bring about the eviction of tenants/owners of premises declared a slum if an ulterior reason arose. A landlord who made financial gain and then refused to maintain a building he was prepared to 'write off', would not be as seriously hurt as the tenants who could be evicted. Once a unit has been declared a slum, further occupancy can be prevented. Declaration of a slum is not invalidated if the occupants, be they owners or tenants, have no alternative accommodation. Recently the Deputy Minister of Constitutional Development was reported to have confirmed that the government would not be responsible for providing alternative housing for Black people evicted from White Group Areas.

Although the local authority has the responsibility of providing housing for the victims of slum removals the case could arise where the local authority simply declares that there is no alternative housing in the White areas within its jurisdiction. The victims, if ruthless use was to be made of the Slums Act, would not have much recourse to normal law. They would have to appeal to the Slums Clearance Court comprised of a magistrate and two ministerial appointees. If the Slums Court found against Black appellants and if the local authorities were 'unable' to offer alternative accommodation the appellants might through force of circumstance be compelled to take up squatter accommodation. Ruthless application of the Squatters Act could then drive the appellants out of the White urban area.

Overview of the scenarios

Even today it is not possible to sketch only a single scenario for the metropolitan areas under continuing conditions of apartheid and the use of three scenarios appears justified because of the uncertainty of which way the government will move in the future. The three-scenario system unfortunately makes consideration of a specific city example rather difficult. Ultimately (probably by the late 1990s), all three scenarios draw a gloomy picture for those who are either currently excluded, or might be evicted, from the White metropolitan areas. If the major South African cities continue to grow and 'develop' within the existing apartheid system then the potential for fuelling violent reaction will be increased and this will be particularly noticeable during the early 1990s.

A SPECIFIC CITY EXAMPLE: JOHANNESBURG

In order to illustrate some of the major points set out in general terms in the above scenarios attention now turns to the city of Johannesburg. For the purposes of the discussion that follows Johannesburg will be assumed to comprise Soweto, Lenasia, and the municipal area of Johannesburg (see Figure 3.1). The discussion here is based on the assumption that the reader has already worked through Scenarios A,

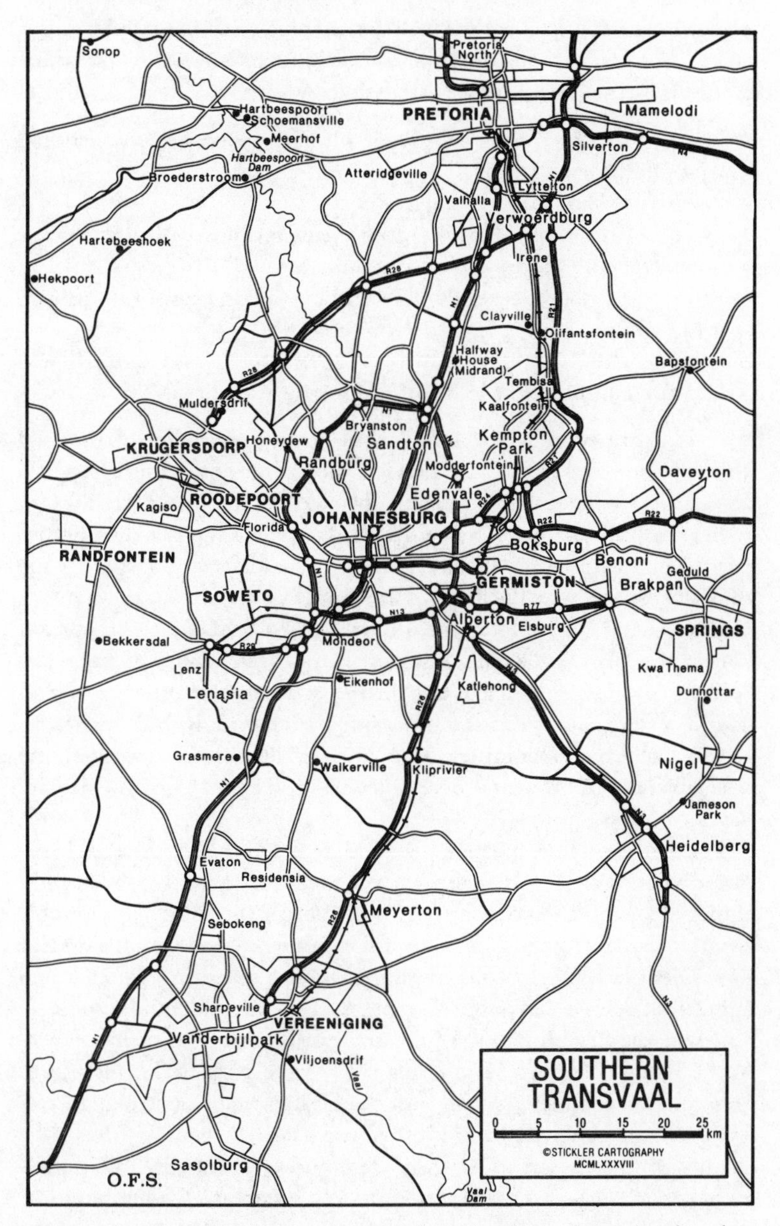

Figure 3.1 Johannesburg, Soweto, and Lenasia within the regional setting of the Southern Transvaal

B, and C. For the sake of brevity the example of Johannesburg under a continuation of apartheid will be presented without following a strict adherence to each of the scenarios in turn. The following points from the main thread of the discussion:

1 What is the present structure of the city in terms of functional zones and racial areas?
2 Where are the main grey areas?
3 What is the character of the grey areas and how will this change under varying modes of continuing apartheid?
4 What is the potential for conflict and in which parts of the city is it greatest?
5 What is the price that will have to be paid for evicting Black/Coloured/Indian people from the present grey areas?

It is assumed that the reader has some knowledge of the real Johannesburg metropolitan area. It should be noted that the CBD remains the single most accessible focus to the built-up area referred to here as Johannesburg. The CBD is also the locus of a large number of job opportunities for Black people and is a focus for a large volume of Black spending. Because of the massive capital investment in buildings and real estate, the Johannesburg CBD is an important cornerstone in the whole assessment rates structure of the city. Around the CBD, particularly on the southern, south-eastern and south-western sides, lies the inner industrial zone with a very large number of job opportunities. The inner residential districts, with the exception of Hillbrow and Berea, are essentially low-rise, low-density areas.

For the sake of perspective it should be noted that if all racial legislation were abandoned overnight the legacy of what has been built under segregation and apartheid will remain for some considerable length of time. Certainly Soweto will continue to exist and the exodus of residents from it to Johannesburg will not result in a sudden and significant decrease of population pressure on its residential resources. Likewise the 'traditionally' White areas will continue to be predominantly White-occupied. The other side of the coin is the picture of Johannesburg under the continuance of apartheid, be it benign, re-enforced, or achieved in any other way, all of which will have the greatest initial impact in the grey areas and adjacent neighbourhoods rather than on Soweto and the 'pure White' areas of Johannesburg. The grey areas of Johannesburg will be the focal points in the discussion which follows.

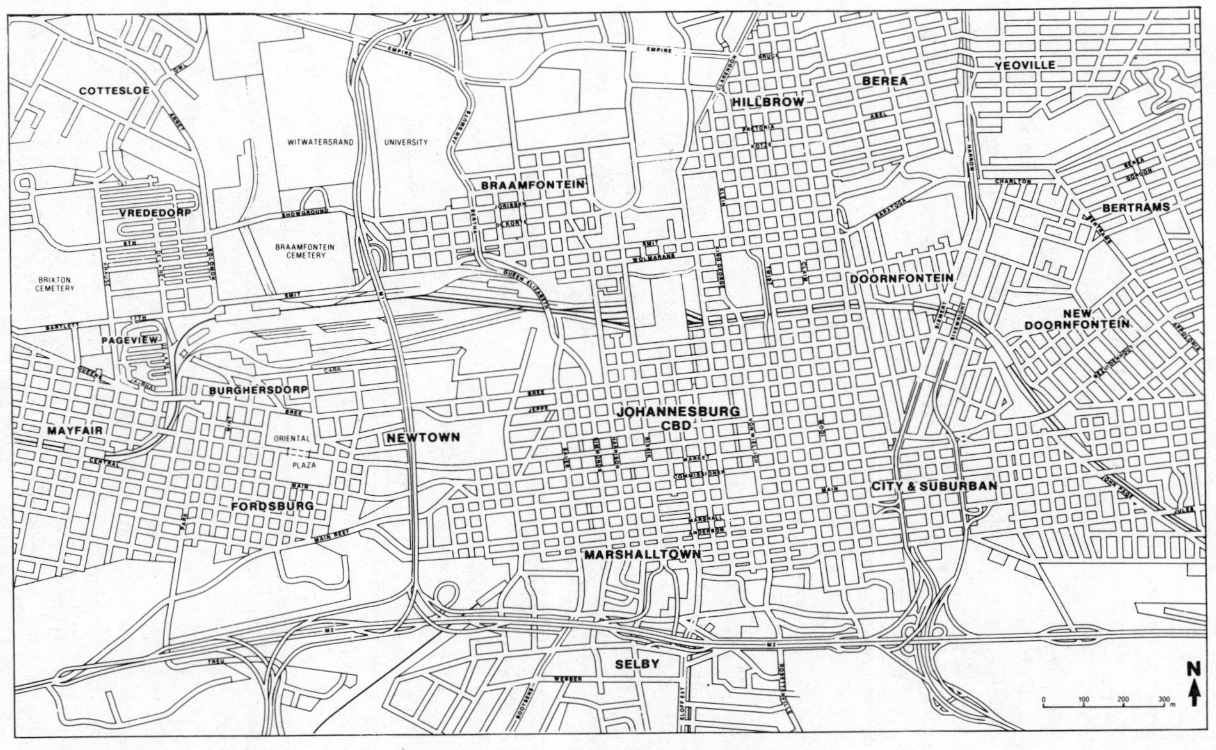

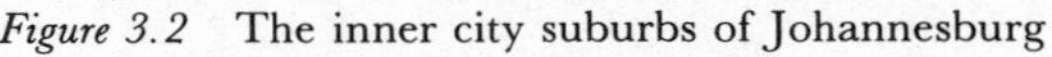
Figure 3.2 The inner city suburbs of Johannesburg

The grey areas of Johannesburg

Although Black/Coloured/Indian people may be found living in many different parts of the city there are three sections of Johannesburg that are frequently termed grey areas (see Figure 3.2) and include the suburbs of Mayfair, Hillbrow, Berea, Yeoville, Bertrams, and Lorentzville. Within the suburbs listed it is possible to demarcate three specific grey areas (Mayfair, Hillbrow, and Bertrams) for discussion purposes. Each of the three identified grey areas will be examined briefly in respect of the impact that continued urban apartheid – as sketched in Scenarios A, B, and C – is likely to have on the residents who have moved into the grey areas from peripheral segregated areas.

Mayfair

The suburb of Mayfair has long been an area of lower-income White working-class people. Over the last six years or so middle- and upper-income Indian families have been moving into the suburb. Although Indian housing is now found in many parts of Mayfair it is most conspicuous in the areas to the south of the railway line and east of Church Street. Indian families have in many cases (if not all) resorted to renovating and improving the properties that they now occupy. Often the original buildings have been totally demolished and a completely new dwelling erected on the property. In numerous instances the renovated/newly constructed dwellings are double storey buildings. The overall effect of the renewal brought about by the enterprise of the Indian people has been to improve the townscape of the suburb and to increase the residential density in an orderly manner, also significantly increasing the capital investment in the area. The presence of a private convent school has greatly assisted in alleviating the problem of finding a proximate school for some of the Indian children.

Under Scenario A it is conceivable that the present Indian sector of Mayfair will continue to expand until such time as a reaction/clash occurs between the Indian residents and the Whites in the area. It must be recognized that under this strategy a clash of some kind is virtually inevitable. The reaction of the government will then probably take the form of requiring permits for Indians who in future wish to take up residence in the suburb and such applications will frequently, if not always, be refused. The question of whether 'permitted' Indians will be entitled to bequeath their properties to their heirs will also arise.

Under the re-enforced apartheid of Scenario B it is possible that future moves by Indians into Mayfair will be stopped altogether and eviction notices will be served on the present residents. Such action, if it occurs at all, will be apparent sooner rather than later and will evince a strong reaction from the Indian residents already in Mayfair. As the Indian community of Mayfair includes businessmen and professional people they will, as a community, almost certainly possess the financial clout to take the government to court. It is not possible to predict the outcome of a legal battle but should it result in the eviction of the present Indian community the cost to Johannesburg will be considerable – not least in terms of eroded community relations. It has been estimated and reported in the press that the potential investment loss in property and buildings could be R50m.

In Scenario C, if the government chooses to use the provisions of the Slums and Squatters Acts for nefarious purposes, then the present Indian community of Mayfair will probably not be harmed. Certainly the quality of the renewal and maintenance of Indian properties in Mayfair is such that there is no likelihood of the area being declared a slum. The only possible exception is related to the fact that extended Indian families frequently live under one roof. Provisions in the local by-laws may allow mischievous officials to believe that grounds for eviction could be sustained.

It appears that the Indian community currently resident in Mayfair is likely to remain in Mayfair and the only questions that arise are whether the community will grow in numbers, spread further into Mayfair, and whether they will have rights of bequeathing property to their heirs.

Hillbrow

Unlike Mayfair, the grey area of Hillbrow which extends eastward into Berea and Yeoville, is characterized by residents from all the main race groups of South African. There are Blacks, Coloureds, Indians, and Whites all living, working, and enjoying the recreational amenities of Hillbrow. The fact that Hillbrow has long been a cosmopolitan area has certainly assisted in the greying process.

As yet it is difficult to foretell whether events in Hillbrow will follow the general lines laid out in Scenario A or B. There is some evidence to date of eviction notices having been served on some Black people in the Hillbrow area, and there have been reports of police harrassment. The extent to which either of those events heralds a change in attitude by the government is not clear. That a significant number of White residents in Hillbrow are concerned about the influx of Black/Coloured/Indian people appears to be a fact.

If the blind-eye approach towards Hillbrow continues, or if the re-enforced apartheid system is applied to Hillbrow, the results will be in accordance with the general statements made in the presentation of Scenarios A and B. If the residents who are not White are evicted, then there will be considerable financial losses for landlords of rented properties. It might be argued that the potential financial losses are so great as to preclude evictions. Yet the fact that one of the major buildings – High Point, occupied mainly by Black residents – belongs to the Anglo-American Corporation might contribute towards the decision to evict tenants! Evictions would create a considerable short-term vacuum in Hillbrow (estimated at 10,000 housing units), and given the depressed state of the economy it might be some time before the vacant premises could be filled. Indeed the mere existence of the vacuum in an area increasingly perceived by Whites to be seedy and unsafe, might deter replacement tenants from filling the empty space. The existence of the vacuum could contribute to a downward spiral in the perceived and actual quality of the Hillbrow environment and the possibility that homeless unemployed people and waifs will break in and camp in vacant premises must be considered high. It follows that the longer there are large numbers of vacant housing units in Hillbrow the less likely it is that a sudden surge of replacement tenants will take place.

A more interesting point would arise if the government were to implement the actions set out in Scenario C. Given the seediness of some parts of Hillbrow at the present time and given the age of a large number of the higher density apartment blocks, it is quite conceivable that slum conditions could arise, or readily be seen to have arisen. Under such circumstances the Hill and environs offers considerable potential for evictions in the 1990s in accordance with what might be termed a 'sanitation syndrome'. With the exception of violence of the kind associated with evictions, and as discussed under Scenarios A and B, the cosmopolitan character of the Hill will probably preclude the type of violent clash that might occur in Mayfair under Scenario A – unless vigilante forces are bussed in.

If the assumption is made that a government could be mischievous in the extreme, then the possibility exists that Hillbrow could be deliberately chosen as a test area for assessing the impact of the formal scrapping of the Group Areas Act. The Slums Act and the Squatters Act would remain in force. The potential for 'normal' old, high density, inner city residential decline is heightened, deterioration actually occurs, and the test is declared a failure. The Group Areas Act is reinstated for Hillbrow. There is a distinct possibility that vigilantes would be bussed in under the above variant of Scenario C.

Bertrams and environs

The grey area of Bertrams and its environs appears to be composed mainly of Whites and Coloureds. Unlike Hillbrow the area is not traditionally cosmopolitan and the existing accommodation tends to be old, medium- to low-density stock. To date there is no evidence, of the kind found in Mayfair, that indicates an upgrading of the housing stock as a result of the greying process.

In terms of Scenarios A and B the medium term will see little change of a kind not already envisaged for Hillbrow. In the longer term, from 1990 onwards, a potential for violent clashes must arise if the Coloured people are still illegal residents and the technikon residences house a large number of conservative and ultra-conservative students. Reports of vigilante action taken by students was briefly reported in the press a year ago. Bertrams is an area which is due for an urban renewal programme. If the programme is introduced without a change in official attitudes then an opportunity to build for the future will have been missed.

Soweto and Lenasia

In concluding this section brief mention is made of Soweto and Lenasia. Consideration must be given to a growing residential inertia brought about by home ownership and improvements in Soweto which will be of consequence with the lifting of urban apartheid.

Under a system of continued urban apartheid a significant number of residents in Soweto who have moved up the income earning ladder will of necessity decide to buy or build their own homes in Soweto. What then occurs is the investment of substantial amounts of capital in an area which in an open city could not command such levels of investment. When the urban apartheid system is disbanded the owners of such properties could well find that they have overcapitalized and will find it difficult to sell the properties in order to enjoy the freedom of an open land market that extends across the whole metropolitan area.

When the Indian community were forced out of downtown Johannesburg, Pageview, and Vrededorp they were encouraged to take up residence in Lenasia. With the passage of years the township of Lenasia has grown into a well-established Indian 'town' with many well-designed and expensive homes. The overall urban environment compares favourably with that of many White suburbs of Johannesburg and is in stark contrast to that which typifies Soweto. It seems reasonable to expect that Lenasia will continue to develop, even under a continuance of apartheid, as an attractive 'town' with a variety of housing types and qualities emerging as a response to the local market conditions. Recently stands and houses at Zakariyya park south-east of Lenasia were put on the market and in two months R2.1m had been invested in the development. Nevertheless, in a post-apartheid society many Indian people will be attracted to properties in the former Whites-only suburbs of Johannesburg: properties that combine quality living with proximity to the business areas of the metropolis. As such Lenasia in general, and residents of Lenasia in particular, will experience the inertia effects of the same kind as those described for Soweto but to a lesser degree.

Attention now turns to how the present metropolitan areas could be transformed.

THE METROPOLITAN AREAS IN A POST-APARTHEID SOCIETY

The discussion below is based on the following assumptions:

1 That the Group Areas Act will be repealed.
2 That the Slums and Squatters Acts will be modified to protect the victims.
3 That all other laws that contain elements of racial discrimination will be scrapped.
4 That there is general political harmony in the country and the leaders are in agreement that all should work towards an improved urban environment.

Given these assumptions it is possible to concentrate here solely on how the metropolitan areas can be improved and to give some indication of what the 'vision' might look like. In order to improve the South African urban society it will be necessary to resort to social programmes as well as demolition and rebuilding programmes in the cities. The concern here is almost entirely with the latter programmes but the former are the key to the success of the whole. Not least will be the need to improve the quality of education at school level. *De facto* single race schools will probably still exist and the multi-racial schools will have to be actively encouraged. It will be imperative that school children know how and why the segregated townships came about. There will be the need for one population group to admit to the errors of the past and a need for the other groups to accept the apology. These comments notwithstanding, the focus here will be on the physical aspects of the post-apartheid South African city.

For the post-apartheid city to become a desirable urban place for all its people, considerable physical construction and reconstruction will be necessary. In the first instance plans must be made to redress the existing imbalances caused by a century or more of planning and building on the basis of segregation and apartheid. One of the most important tasks that will have to be undertaken is to increase the densities of lower income housing closer to the inner city job opportunities. As indicated in the earlier discussion in this chapter, many South African cities contain inner residential zones that are both ripe for redevelopment and which are zones where population densities can be increased along with substantial improvements to communal amenities. Care will have to be exercised in the planning process to avoid simply building new potential slums such as has occurred in the 'new' Pageview. Higher density living can be achieved in the inner areas in conjunction with the creation of more open space, disregarding much of the original city grid of street blocks and avoiding the construction of massive residential tower blocks for low income residents. Two world examples which stand out, and are worth closer scrutiny for both their successes and failures, are the renewal of the inner residential zones and CBD of Toronto, and the provision of working-class housing in post-Franco Madrid.

The 'vision' of South African metropolises is of urban places with thriving central areas which are surrounded by resident populations living in closely packed clusters of medium-rise inner city residential units: apartment blocks of four to six storeys, and double or treble storeyed cluster housing. It is essential that the buildings should not be as tall as those which currently exist in Hillbrow and at the same time the buildings must be designed to increase the population densities: a *de jure* height limitation and an acceptably high *minimum* bulk factor are required. The inner city residential areas would be designed to create a sense of community with open spaces in the form of parks, open playgrounds, sports courts, and greenways serving as central focuses for individual residential districts and sub-districts. the planning and design will be such that a range of socio-economic groups will find the inner city areas as attractive as is now the case in many European cities. The vision is one of upper, middle, and lower income residents of all colours and all ages living in the innermost residential zone (1 to 2 km from the centre of the CBD) where there will be clusters of accommodation that cater for varying demands of utilitarian and luxury living. All residential units will contain adequate amounts of off street parking for private motor vehicles. Proximity to place of work will minimize the use of private transport during working hours. There will be inner city playschools and junior schools with senior schools positioned farther out.

Outside of the above-mentioned innermost zone of residence, office, and retail activities and beyond the inner industrial zone, will lie what is *currently* termed the inner residential zone. In its revamped form the second residential zone will house the main concentrations of working-class people who have jobs in the CBD or in the inner industrial areas. In general the new higher density innermost residential zone will/should extend initially for a radius of 2 km with the second residential zone being between 2 and 4 km from the heart of the CBD but avoiding established inner industrial zones. Densities of the kind envisaged will provide the population base not only for CBDs that are alive both in the daytime and at night but will also provide the base for improved inner city public transport services in the form of buses, roving taxis, minibuses, tube trains, or light rail systems.

If the Group Areas Act and associated legislation were to remain repealed then the 1990s could witness at least some progress towards the 'vision'. Most noticeable would be the continued greying of the existing grey areas along the lines suggested in Scenarios A and B. There could also be some movement of Black/Coloured/Indian people into the relatively higher density suburbs which comprise single storey detached housing, e.g. Parkhurst and Norwood in Johannesburg, and most of the southern suburbs in Cape Town which lie in close proximity to the commuter railway line. Wealthier people might well buy into the high status former Whites-only suburbs. The main difference between the apartheid and post-apartheid scenarios would be the lack of institutionalized evictions and removals. Actual acquisition of properties and construction would probably not take place in any significant way for some years. Given that the Toronto scheme took fifteen years to achieve its final goals, a time span of twenty to thirty years would be reasonable in the South African metropolitan areas.

In the post-apartheid scenario of the mid-1990s some of the problems associated with overcapitalization in the former apartheid-generated peripheral Black/Coloured/Indian townships will have manifested themselves. It is most likely that the major locations/townships, where the majority of Black people currently live, will still be in existence and providing housing for the lowest income earning groups, the Black social pensioners, and many of the 'casual' unemployed of the Black community. Although some Black people will have vacated the former locations they are unlikely to have done so in such numbers as significantly to relieve the housing shortage. Nor will employment opportunities have increased to such an extent as significantly to relieve the unemployment problems in the townships. As a consequence three important steps could be taken at an early stage in the post-apartheid system that would alleviate some of the costs of housing/shelter. In the first instance persons who have continuously rented a house (particularly a 'matchbox' house) in a Black township for a specified time (say 15 or 20 years) must be assumed to have paid off the house and must be able to take ownership. Second, rent defaulters who are unemployed must be afforded more grace than is necessarily the case at present. Third, controlled squatter areas have to be tolerated on the outskirts of the metropolitan areas to serve as a first staging post to new rural migrants who are unable to buy into the housing market. Finally, considerable changes to the methods/systems/means of financing Black low income housing will have to be implemented and a range of alternative approved building materials such as timber framed housing will have to be developed, encouraged, and permitted. In the post-apartheid lowest income areas brick and mortar houses alone will not be sufficient to provide the necessary first levels of shelter.

To conclude the chapter attention turns briefly to the non-metropolitan areas of South Africa.

THE NON-METROPOLITAN AREAS

As indicated in the preamble to the chapter there are several minor urban areas in South Africa which lie between the major metropolitan areas and the non-metropolitan areas. In this section concern is primarily with the towns and minor towns of the hierarchy. Brief mention will be made in passing of the minor metropolitan areas that represent the transition point between the upper and lower ends of the urban spectrum in South Africa.

The non-metropolitan areas in South Africa are characterized by small populations of White people and relatively and absolutely large Black/Coloured/Indian populations. The perceptions of what is attractive and good, and what is unattractive and bad about the non-metropolitan areas will vary sharply depending upon whether one is a White, Coloured, Indian, or Black resident in a small town. The 'Black' populations of the non-metropolitan areas, with few exceptions, are threatened by the same four 'horsemen' who 'ride' through the metropolitan townships – namely: poverty, overcrowding, homelessness, and unemployment. Ironically, in the non-metropolitan areas the plight of the Black residents is made worse by the decline in the numbers of White people who have drifted to the metropolitan areas. Given the apartheid system, it has been the Whites in the small towns who have owned and generated the wealth on which jobs for members of the other population groups depended. With the drift away of the Whites, and the continuance of the apartheid controls, vacant properties and businesses of the departing Whites cannot be taken over by people who are not White and who stay on in the smaller urban places living under conditions that grow worse with the passing years.

Perceptions of the advantages and disadvantages of the non-metropolitan areas

From the point of view of White people living in the urban places of the *platteland* there are a great number of advantages compared to living in the large cities of South Africa. Land and labour are cheap, and these two factors mean that a high standard of living can be realized for many Whites at only moderate costs. The low cost of labour and land will offset the relatively high cost of certain building materials and consumer goods to a significant extent. Once established in a small urban place Whites find that their children are 'protected' from the evils of the city, they enjoy reasonably good to very good schooling in small classes at well-equipped state schools, and they all breathe clean fresh unpolluted country air. In the White communities of the small towns there tends to be a strong community spirit, a sense of belonging, and with the exception of certain places on the northern borders of the country there is a feeling of 'security' largely built around *kerk en saamelewing*.

The advantages experienced by the Whites who live in the non-metropolitan areas more than offset the disadvantages which, if a person is employed or able to make a living, are few in number. In a significant number of small towns children of high-school age will have to leave home to attend a boarding school in a larger place. Young people wishing to attend institutions of tertiary education will almost always have to do so as boarders. For the small town residents there is the need for periodic trips to larger places to shop for higher order consumer items, and to obtain access to professional services that are not available locally.

The perceived advantages for Blacks who live in the townships or *lokasies* of the non-metropolitan areas are few indeed. Apart from apparently living in areas where current township unrest is either non-existent or very infrequent, there can be no advantage over living in a Black area attached to a White metropolitan area, notwithstanding the disadvantages of that style of living which were set out earlier. Certainly there are a number of disadvantages associated with life in a non-metropolitan Black location. Schooling, albeit less plagued by unrest, is generally of lower standard than in the metropolitan areas, the locations are small and very crowded, and in many cases the journey to work from the location to the White town is by foot. In some places there are bus services, but the fares relative to wages earned are high. The quality of housing is certainly no better, and in many cases worse, than that in the metropolitan townships, and air pollution in the *lokasie* is as bad or worse, given the fact that the majority of Black small town locations have no electricity – coal, wood, or dung fires being used for all heating and cooking. Wages earned for tasks done are lower than would be the case in the metropolitan areas, and foodstuffs in the local shops will be relatively and absolutely more expensive than are the same items in the metropolitan areas where economies of scale are greater.

Overall the advantages of living in the non-metropolitan areas lie decidely with the White community rather than the Black community.

Problems associated with the non-metropolitan areas

The single most important problem associated with the non-metropolitan areas, more particularly the smaller urban places, is the process of depopulation. The loss of White people from the small urban places erodes the job opportunities for the other population groups in the place and they in turn are 'encouraged' to drift to the cities where in many instances their first place of abode will be a shack in a squatter camp. Urban Whites who leave the non-metropolitan areas are to an extent the 'victims' of the White rural depopulation process, which in turn is being accelerated by a series of devastating droughts, increased costs of farming, poor farming practices, and the consequent loss/sale of land to large agri-businesses. In short, the erosion of the rural communities which were the *raison d'être* for the urban service centres has precipitated the declining fortunes in many non-metropolitan areas.

Over the last 20 to 30 years a substantial network of tarred roads has been built by the state and these now provide outstanding links to virtually all places in the South African urban system. It could be assumed that the excellent road network would be a positive component in the development of the rural areas and the urban service centres. In fact the opposite is true in many cases. The rural communities who supported local services centres now find it just as easy to travel to larger urban centres in order to purchase their domestic and agricultural requisites. The improved roads have therefore accelerated the rate of decline of many smaller urban service centres.

The non-metropolitan areas under continuing apartheid

The separation of people in the non-metropolitan areas is already far greater than is the case in the metropolitan areas. Although many chain stores, franchised fast food outlets, and entertainment facilities have a non-racial customer/patron policy, the real openness of such facilities in small-town South Africa is not on a par with what occurs in the metropolitan areas. The broad 'education' in race relations, gained from living in a large cosmopolitan city, has not yet reached the White *platteland* towns. Indeed, it is in the rural areas that the belief in White supremacy in all respects is strongest. Therefore if the policy of apartheid remains as it is then the hope for any voluntary sharing of space and facilities in the *platteland* dorps is nil. The only possible exceptions could occur in some of the smaller towns in the Cape province and Natal where the White populations have declined considerably and where as a result an accommodation between Coloureds and Whites or Indians and Whites might be possible. Even a system of re-enforced old-style apartheid, as considered under Scenario B in the discussion of the metropolitan areas, will have little effect on an already deplorable situation of inequality in the social well-being and life-styles of non-metropolitan Blacks/Coloureds/ Indians. In short by the mid-1990s the situation for most non-metropolitan Blacks will be worse than it is now and it will be considerably worse by the year 2000. It can confidently be stated that in many instances small towns will have become ghost towns by then unless the limitations imposed by apartheid are removed, and carefully thought-out upliftment programmes which include financial subsidies and improved access to finance are introduced. Given the lack of natural economic bases and the absence of industry in most parts of the declining rural areas the prospect for saving the smaller urban areas looks grim even if apartheid is scrapped.

The non-metropolitan areas in a post-apartheid society

We now come to the most depressing aspect of this chapter. If all apartheid and racially discriminatory legislation were scrapped, the legacy of apartheid would linger on for a long time in the non-metropolitan areas. The analogy that comes to mind is the differences between the metropolitan areas in the north-eastern parts of the USA and the metropolitan and urban areas in the southern states. Legislative changes alone will not rectify a deplorable situation that has been actively generated over a century or more. It will only be in those small urban places which are virtually abandoned by the Whites that Blacks will be able to 'take over' the former White town. Given the reasons for White depopulation of the rural areas and the small urban service centres, the Blacks will largely inherit urban places that have lost their economic base.

In the large urban places and minor metropolitan areas such as Bethlehem, Kroonstad, Nelspruit, Pietersburg, Newcastle, Kimberley, Beaufort West, Bloemfontein, Potchefstroom, and Klerksdorp, the possibilities of racial clashes in a post-apartheid era, created by the scrapping of racial laws but without a change in attitudes, will be high. In brief the possibilities for a harmonious and prosperous future for the non-metropolitan areas even after apartheid is scrapped looks bleak. In those non-metropolitan areas where the Black population is not organized into, and by, social movements and unions, the possibility of post-apartheid racial clashes will be lower than will be the case in places such as Bloemfontein, Welkom, Newcastle, and Kimberley where Black social movements will exert more pressure in order to exercise their rights gained from the scrapping of apartheid legislation. The spark for such clashes need not come from the Black community but would most likely come from the conservative and ultra-conservative White population groupings: the approximation is with the vigilante scenarios discussed for the major metropolitan areas.

CONCLUDING STATEMENT

Proceeding from an appreciation of the existing state of affairs in both the metropolitan and non-metropolitan areas, assessments of what the future urban places will be like over the next decade have been presented. In the case of the metropolitan areas, the views of the future were couched against a backdrop of three scenarios. For the non-metropolitan areas the presentation was more general, as the possibility of formal or informal reform as it might effect the residential component of urban places over the next decade is extremely low. In short, a significant amount of residential and commercial reform is possible in the major metropolitan areas by the mid-1990s; the same might be possible in the smallest urban places which have been almost totally vacated by Whites, and significant residential and commercial reform will be virtually nil in the middle order towns and minor metropolitan areas which are at the heart of the *platteland*.

Finally, the key to a harmonious and prosperous urban living in South Africa depends as much upon a sound schooling for all as upon the political accommodation of all its peoples. It is of fundamental importance, therefore, that the gross inequalities in the racial education system are removed. A significant number, and thereafter a growing number of multi-racial schools need to be established and encouraged. The school curricula will have to include revised syllabuses in social science subjects such as history and geography in which the truth about the origins and nature of the real problems of South Africa are presented. Only from such a base might future generations of South Africans be in a position to accept a common challenge to dismantle the legacy of apartheid which will linger for many years after it is formally scrapped.

Note

This chapter was in press when the Group Areas Act was rescinded. However, the fragile nature of change in South Africa and the unpredictable reactions to whatever situation may evolve, ensure that the scenarios offered by the author retain their validity.

4

URBANIZATION AND URBAN SOCIAL CHANGE IN ZIMBABWE

David Drakakis-Smith

INTRODUCTION

The aim of this chapter is to examine the nature of the socialist transformation in Zimbabwe in so far as this affects the urban poor. In order to achieve this objective, the chapter will commence with an historical review of the development of the state, particularly with respect to the urbanization process. As will become evident, it is not possible to understand the present contradictions in the country without reference to its short but intense colonial past. The chapter then examines the recent changes in government policy since independence was achieved in 1980. In particular, it focuses on policies and programmes directed towards the urban poor in order to assess the ways in which they have benefited and might expect to benefit from the accession to power of an avowedly socialist government. This assessment will primarily be undertaken by means of a review of the urban supply systems for two basic needs – namely, food and housing.

Although a small, less developed country by world standards, Zimbabwe comprises one of the more broadly based economies of southern and central Africa. It not only exports a wide range of primary products, but also manufactures a considerable amount of consumer and capital goods (see Stoneman 1984; Thompson 1984). This economic strength could provide an important regional base for lessening the dependence on South Africa, but the Zimbabwean economy itself is still very much controlled by expatriate enterprises and overseas capital, much of it South African.

Within Zimbabwe this expatriate dominance still has a marked spatial dimension, despite a decade of independence, which derives from the distribution of land during the colonial period. This is not to imply that a dual economy exists in Zimbabwe. Although production and reproduction appear to be fairly distinct in socio-spatial terms, there always has been and continues to be a strong relationship between black and white, rural and urban, peasant and capitalist sectors of the political economy which is most manifest in the transfer of value through labour exploitation. The situation is further complicated in Zimbabwe by the division and conflict within expatriate capitalism between settler and neo-colonial interests. The nature of this conflict has strongly influenced the spatial dimensions of urban and economic development.

Since independence the 'growth with equity' strategy of the socialist government has been threatened by severe developmental problems. Some of these stem from the erratic nature of the economic growth that has occurred since independence in 1980. In part this is the consequence of cumulative external factors, such as the global recession and the rising cost of fuel, but internal factors have also contributed to the overall difficulties facing the socialist government. Amongst the most prominent of these is the very rapid population growth, currently standing at 3.6 per cent per annum (World Bank 1989).

This massive growth rate has brought about severe pressure on individual households, particularly in the communal lands where subsistence production is increasingly falling short of basic requirements. With the gradual relaxation of restrictive legislation, this has consequently led to accelerated urban migration in search of cash supplements to family incomes; a trend which some claim has been knowingly fostered by government policy (Simon 1986b). The socialist government is aware of these new socio-spatial shifts and is attempting to deal with them in its development programme. As in many other African countries, the emphasis is on rural redevelopment. This is perhaps not surprising given the extent or rural poverty in the tribal areas of Zimbabwe, but it also reflects general tendencies within African socialism. However, given the relative recency of Zimbabwe's independence, the opening up of new opportunities for international capitalism, and the guarantees to white settler capitalism that the Lancaster House agreement forced upon the new Zimbabwean government, it is perhaps not surprising that socialist influences in development planning, particularly as they affect urban growth, are difficult to identify. Since 1980 a complex mix of forces has shaped development and it would be counterproductive to examine them merely within the apparent dualistic framework of a black, rural, peasant and a white, urban industrial economy. It is to an identification of these forces that we now turn.

URBANIZATION DURING THE COLONIAL PERIOD

The urban system in Zimbabwe is a creation of settler colonialism. It functioned primarily in a compradore capacity to facilitate the export of various primary commodities and the import of consumer goods, is located in the former white areas of the country and has always accommodated the majority of the white population whose urban proportion has steadily risen from 52 per cent in 1911 to over 80 per cent today.

From the beginning there was a small urban proletariat but for the most part this comprised white workers who, attracted by high wages, identified themselves primarily with the agrarian settler class rather than the non-settler capitalists, and certainly not with their black, urban counterparts. Not that there were many black employees in industry or tertiary activities. There were, however, large numbers of black domestic servants in the urban centres whose incorporation into the economy was organized in the same way as farm labour. Their residence in the white towns was tolerated only so long as they had a job (and therefore accommodation) and their families were not permitted to move from the (TTLs).

Changes first began to appear in the settler political economy during the Second World War when the growing demand for semi-processed raw materials, and the curtailing of external supplies of consumer goods began to stimulate domestic manufacturing. This was further encouraged by the creation of increased 'domestic' markets in southern Africa, particularly during the short-lived Central Africa Federation (CAF), and the growing availability of foreign investment funds. The result was an urban-based growth of overseas manufacturing capital to challenge the domination of the white agrarian bourgeoisie (Ndhlovu 1983).

The period from 1950 to the unilateral declaration of independence by Southern Rhodesia in 1965 was marked by a series of contradictions and conflicts in the role of the state in Zimbabwe. On the one hand, labour movement into the cities was 'encouraged' by legislation which limited the size of agricultrual holdings; on the other hand, the growth of 'surplus' labour in the cities was actively discouraged by vigorous enforcement of the vagrancy laws. Ironically, although blacks had long outnumbered whites in the towns (usually in the order of 2.1), about half of these were non-nationals from nearby states, a trend that continued to grow until the 1960s (Potts and Mutambirwa 1989).

In many ways, however, circumstances were changing irrespective of the wishes of the white agrarian bourgeoisie that formed the 'centre of gravity' of the class structure (Arrighi 1973). Decades of increasing population pressure on the TTLs had led to a relative decline in subsistence production so that by the 1960s landlessness had become a feature of the poorest areas. As a result, black migration to the main cities was beginning to expand and, despite the wishes of the municipal authorities, squatting and 'lodging' (sub-letting) was appearing in the more densely populated African townships of the capital (Nangati 1982), with the growing number of urban unemployed apparently being used to keep urban wages low (Davies 1988). Given the doubling in size of the black urban workforce since the 1930s, it is not surprising that such conditions led to increasing black militancy (Davis and Dopcke 1987).

Nevertheless, attracted by the prospect of the CAF, increasing manufacturing capital was beginning to challenge some of the basic principles of settler capitalism and class structure by permitting the development of a black urban proletariat, by the recognition of African trade unions, and by encouraging the emergence of a black petit bourgeoisie (Baylies 1981). Such changes were to strike an increasingly sympathetic chord in some government circles, but many traditional settlers felt that it was inadvisable. Certainly few municipal authorities felt this way and envisaged little material change. Thus, living conditions for many urban blacks were still poor, subject to harrassment, and generally insecure. Eventually settler capitalism perceived a threat to its political domination from this alliance of various nascent African urban classes and manufacturing capital, much of which was a complex mixture of British and South African. As a result the main settler classes – namely, the agrarian and urban bourgeoisies and the white urban proletariat (reinforced by recent 'assisted migrants') – closed ranks and declared UDI to preserve their racial superiority in the political economy.

In reality, despite the legislative intent of these and other measures, the UDI period witnessed a continued urban movement of Africans, although the urban proportion of the total African population remained at 14–15 per cent. The process was gradually intensified by the steady mechanization of many of the larger commercial farms; a change which constitutes a virtual admission of the redundancy of settler agrarian principles; the economic bases of which were maximum appropriation of labour surplus and minimum reinvestment. Rural wages thus remained very low, and poverty increased.

During UDI the manufacturing contribution to GDP (mostly consumer goods) rose from 10 per cent to 24 per cent (Stoneman 1979). The UDI government thus became increasingly dependent on surplus appropriated through manufacturing and/or foreign-owned production units. Given also that major sections of settler capitalism were export in orientation, it is not surprising that the settler state did little actively to discourage the growth of neo-colonialist forces within the country. In one sense, therefore, UDI can be seen as a temporary delay on the road to neo-colonialism. The black urban population, largely as proletariat, has continued to grow. In common with many other developing countries, despite the fact that many individual leaders received their political education in the cities (such as Nkomo and Ndhlkova in the trade union movement), the liberation struggle was rural-based and the swollen urban proletariat did not take a major role in the independence struggle. What had occurred, however, was a spatial shift in the reproduction of labour. Whereas the predominantly male urban labour-force was formerly reproduced through the subsistence activities of the peasant mode of production, it is now reproduced through an urban subsistence role performed partially by the non-proletarian members of the household, usually the women.

It is doubtful whether the neo-colonial reformists within the state envisaged the spatial changes in which they were involved as clearly as this. Although a fairly elaborate planning structure existed in pre-independent Zimbabwe, regional planning primarily consisted of 'accepting' that the urban growth throughout the European areas would continue to occur and should therefore be 'planned' to avoid excessive concentration in Salisbury and Bulawayo. Most of the urban planning *per se* was physical and designed to cope with demands for accommodation, whilst maintaining segregationalist principles (see Wekwete 1987a; Teedon 1990; Rakodi this volume).

The provision of urban social services under colonialism thus continued to reflect both racial and class cleavages. This favouring of the white population meant that social service provision was urban-biased. Thus undernutrition and malnutrition were far worse in the rural areas than in the larger cities. Similarly, health-care provision was strongly structured towards urban-based hospitals or clinics. However, it would be incorrect to assume that this urban bias in the provision of social services extended to the black population. Certainly there appears to have been some benefit from living in the city with the infant mortality rate amongst urban blacks being much lower than for their rural counterparts but nowhere near as low as for the white population (Loewenson and Saunders 1988). Furthermore, even the limited social welfare provision for blacks was paid for by the Africans themselves through levies on beer sales in the townships (Musekiwa 1989).

Thus, despite the ostensible lack of coordinated social planning in Rhodesia, the reality of ninety years of colonialism was a clear spatial dimension to the class structure and modes of production across the divisions of lands and the location of urban systems. As in most Third World countries the modified pre-capitalist mode of production was predominantly rural, and the imperialist/neo-capitalist mode of production was predominantly urban. In addition, there was a third mode of production: that of the settlers, which was predominantly rural and which had evolved and was maintained through the manipulation of space. A further consequence of the control of movement between these social spaces and the regulation of the cities was the virtual absence of a petty commodity sector. All this was reflected in the provision of urban social services.

POST-INDEPENDENCE URBANIZATION

The post-independence period has witnessed a continuation of the spatial and urbanization processes which emerged during the 1970s under the liberal reformist banner – accelerated by the rescinding of legislation restricting African movement, and residence and property ownership. The combined weight of settler and international class interests at the Lancaster House independence negotiations has clearly muted the impact of the socialist government in changing the social relations of production. This is nowhere more evident than in the redistribution of agricultural land which, although much has been achieved, still reflects the patterns formalized half a century ago (for details see Zinyama 1982, 1986). Indeed, for many, rural poverty has increased since independence due to the stagnation of employment opportunities on commercial farms, the slowness in land redistribution, and a continuing high birth rate (Mutizwa-Mangiza 1986; Weiner 1988); the result has been a notable rise in migration to the urban areas.

The movement into the cities has accelerated for the same reasons that it has in almost all newly independent Third World countries over the last twenty-five years – namely, the removal of restrictive legislation and the promise of employment, either in industry or in government positions no longer the exclusive prerogative of expatriates (Potts 1987). As elsewhere, however, the pace of the influx has exceeded the capacity of the cities to absorb large numbers of low-income migrants: in terms of jobs, housing, and many other aspects.

Although national data are difficult to put together, it would appear that the already dominant cities of Harare (the capital) and Bulawayo have received the main impact of this in-migration (Simon 1986b; Potts 1987). The current population of Greater Harare (which includes the nearby satellite town of Chitungwiza) is well over 1.5 million; it comprises more than half of the total urban population and is predicted to double by the end of the century (Chikowore 1989). Clear evidence of the rapid growth of the capital is the unprecedented appearance of large squatter settlements on the edge of the city. The biggest of these at Epworth grew from 5,000 to 30,000 in 1982 alone, and currently stands at some 50,000. The numbers of households living as 'lodgers' (sub-tenants) in the low income areas of the city has also increased rapidly since independence (Horrell 1981; Teedon 1990; Phiri 1990).

The growth in the urban population has also brought about changes in the composition of the labour force. In 1969, as a result of employers offering hostel accommodation only to men, the male urban population was almost double that of women. As legislation permitting ownership and family residence was relaxed and removed, so virtual gender parity has been reached. However, bias is still clearly evident in the patterns of employment. Only about 15 per cent of the non-agricultural workforce is female, even in Harare (Drakakis-Smith 1985).

Although this proportion does not appear to have risen much over the last decade, once again the aggregate figures mask a considerable change. In absolute terms the number of women entering formal wage employment in the non-agricultural sector has almost doubled during the last five years to more than 100,000. Within this overall growth there has been a major switch in occupational patterns. Most significant has been the massive fall in the importance of domestic service (from one-half to one-fifth of all waged women) – a fall which is likely to continue with the flight of wealthy white families and the introduction of further socialist labour legislation in minimum wages.

The fall in female domestic employment has not been accompanied by a switch to manufacturing. In 1982, of the 76,500 employed in this category in Harare only 6,000 were women. The principal shift appears to be to the distributive trades, which is primarily licensed self-employment in petty commodity trading in food or knotted/crocheted articles. However, as previously noted, one of the principal roles of women in the urban context has been to lower the cost of labour by supplementing the domestic costs of reproduction through non-remunerative but economically valuable (and valorizable) activities such as urban gardening. It is clear, therefore, that the increased freedom given to family settlement in the cities has lowered labour costs to manufacturing capital by encouraging the development of urban subsistence and petty commodity sectors, primarily by women (Drakakis-Smith 1985). It is in this context that the social changes engendered by independence have been somewhat disappointing for urban women (Batezat *et al.* 1988).

SOCIALIST POLICIES TOWARDS URBAN SOCIAL CHANGE

There has been considerable debate on the development options open to Zimbabwe following independence (see Ndhlovu 1983; Cokorinos 1984, and Davies 1988 for useful summaries). Some argue in favour of a reformist strategy which leaves much of the economy under the control of the private sector whilst the state redistributes wealth through a 'basic needs' strategy. Other, more radical voices, call for a much more powerful role for the state in controlling economic resources and directing development strategies.

The course of events since 1980 has shown that a complex mix of forces has been in operation, most of them persisting from the country's colonial past. As might be expected, these have generally moderated the more radical development strategies, particularly as ZANU-PF has sought to 'establish its credentials' both domestically and internationally. In this context, the retention of much of the colonial bureaucratic system together with a surprisingly substantial minority of white professionals, managers, and administrative staff (Sibanda 1988) has played a major role in maintaining both settler and international capitalist influence, not least through loan and aid schemes aimed at the continuation of economic stability and which themselves have introduced international bureaucrats into the administrative system. Moreover, the increasing incorporation of black Africans into the state bureaucracy has also served to blunt the development of radical politics by encouraging the emergence of a black bourgeoisie (Schatzberg 1984). In short, a conservative filter exists between policy-makers and the masses.

With reference to the development of urban social programmes, several broad background factors must be considered before specific policies can be considered. One of the most important of these is the ideological commitment of ZANU-PF to rural development through restructured production units. This philosophy is shared by neighbouring socialist governments, but unlike its neighbours Zimbabwe has inherited a comparatively well-developed urban industrial sector that it is hoped will play an important role in the development of economic independence from South Africa on the part of the SADCC group of 'front-line' states. On the other hand, many members of the government feel that the black urban proletariat did not make any significant contribution to the liberation struggle and, indeed, proceeded to benefit from the expansion of manufacturing during UDI.

There are those, both inside and outside the socialist movement, who argue that uncontrolled enthusiasm for rural investment at the expense of the urban-industrial sector may well undermine this national advantage (see Yates 1981). Such arguments have been underlined by the 'failures' of the Maoist/African socialism of Nyerere and Machel and their enforced retreat into a 'pragmatic socialism' acceptable to the donors of international aid (Weaver and Kronemer 1981; Zafiris 1982).

A second major factor influencing the nature of Zimbabwean socialism is related to this point, and concerns the degree to which the state's prosperity is still dependent on capitalist rather than state-controlled agriculture and industry – not only in terms of export earnings but also in terms of the national employment structure. Almost 60 per cent of the black workers in Zimbabwe are employed in occupations dominated by private sector employers (usually white), both domestic and international. In this context, of course, Mugabe was bound by the Lancaster House agreements to guarantee compensation for displaced or nationalized private/white enterprises, but he has frequently expressed a desire to see the emergence of more black entrepreneurs and enterprises. One result of this has been the rapid incorporation and growth of a black bourgeoisie.

Given such constraints, it is perhaps not surprising that socialist policies directed towards resolving the problems of socio-spatial inequality in Zimbabwe have been limited in both their scale and intensity. In addition, spatial development has been considerably affected by another ethnic factor which has been in existence much longer than the present state – namely, inter-tribal rivalry. The traditional antagonisms were suppressed during the independence struggle, but subsequently resurfaced in civil disorders in the southwest. Although there is little evidence as yet to indicate that these disturbances were linked to developmental discrimination (and perhaps more to indicate a South African involvement), it is certainly a factor which cannot be discounted within policy formation.

So, within these ideological, constitutional, ethnic and economic constraints, what are the policies of the socialist government with respect to alleviating regional and urban inequality and what has been achieved since independence? It must be stated at the outset that as yet no national or comprehensive urban development strategy has emerged from the socialist government. The initial policy document covering urbanization (Riddell 1981) was essentially an aspatial, sectoral statement although it, nevertheless, had considerable spatial implications – not least through the recommendation to unify families split by migration by bringing the rural segments into the city.

As Carole Rakodi clearly illustrates in Chapter 5, the framework for urban planning is a legacy of the colonial and UDI periods which means that it has a reasonable capacity for local autonomy and delivery. However, since independence both the fiscal base for individual towns and the rationale behind the planning process have been weak and the prevailing attitude seems to be one of coping with ever-increasing needs on an *ad hoc* basis.

New urban planning principles have slowly begun to emerge during the 1980s. Worthy of note is the much greater emphasis being given to the development of smaller towns, largely as service centres for the neglected rural areas, through a seven-tier hierarchy (see Wekwete 1987a). However, development is as yet proceeding slowly and has failed to stem the rapid shift of population to the larger urban centres such as Harare and Bulawayo. Indeed, one senior minister recently identified as a major planning objective 'the creation of a significant proportion of new jobs in Harare so that the city remains the most desired destination [for] job-seekers' (Chikowore 1989). More recently the five-year plan (1986–90) outlined a series of objectives for urban development and these have led to the establishment of an urban development corporation (UDCORP) the role of which is to assist and stimulate development in designated growth points (Wekwete 1987b). As yet, however, the brief of urban planning seems to be to stimulate urban economic growth and there is little sign of a national urbanization policy which sets out a social programme which may be integrated with economic growth. A recent government statement on the planning priorities for Harare's development placed little direct emphasis on social needs in the capital (Chikowore 1989). It is to this specific topic that we now turn.

One might justifiably pose the question as to what is to be expected of a socialist government with regard to urban social programmes. Indeed, it may be more apposite to ask what is expected of a socialist government in transition. Davies (1988: 22) claims that the role of the state is crucial not only in intervening on behalf of the poor but also in promoting 'their development into a self-conscious revolutionary class'. He argues that unless this objective is paramount it is difficult to distinguish any socialist programmes from social-democratic intervention on behalf of the poor. In theory, this gives us two sets of criteria against which to assess urban social programmes. First, as a means towards helping the oppressed 'to create a self-conscious revolutionary class' who will be the vanguard of future socialist change; and second, as part of a more general democratic attempt to help 'the poor' to improve their life-style by giving them access to social resources.

As will soon be evident, analysis along the former lines is as yet unrewarding since it is very difficult to distinguish socialist objectives within the social welfare programme. On the other hand, we must not be so unaware as to believe that social improvement will be widespread or effective without a redistribution of wealth. For it is poverty above all that continues to make even the most social democratic of objectives difficult to realize.

There were great hopes that the introduction of minimum wage levels would be a major step forward in promoting greater equality in development. As always, however, minimum levels tend to become standard rates and although they have proved beneficial for some rural workers and domestics (many others were threatened with dismissal), they are pitched far too low for industrial workers and it is alleged that some multinational firms have been able to reduce wages (Nyawo and Rich 1981: 91). Given the failure of the government to control the cost of basic commodities, it is estimated that in real terms wages are now lower than in 1979.

As noted above, the present government displays no great affection for urban workers, particularly their trade unions. Much of this distrust stems from the pre-UDI period when trade unions were dominated by the white and black middle class and were used by the colonial government as industrial peace keepers. This suspicion has persisted through to the present (see Cheater 1988) although ironically the government now regards strikes as not being in the best interests of the state. The government still regards urban wage earners as privileged compared to peasants whilst, for their part, union leaders feel threatened by what they see as attempts to undermine their authority amongst the proletariat by the establishment of workers' committees whose primary loyalty lies with the party rather than the proletariat.

In fact, Cheater (1988) feels that the urban proletariat in Zimbabwe has very little solidarity of interest and action. Notwithstanding the rivalry between union membership and workers' committees, it appears that workers prefer to rely on the 'moral community of the industrial enterprise rather than on government or their own class solidarity' (Cheater 1988: 297). Significantly, workers promoted to supervisors are seen as moving towards the interests of white management rather than capital *per se*, and yet are used as brokers for workers' interests.

It is within this context of mutual suspicion and interdependence between the urban workforce and the socialist government that any analysis of the fragmented social programmes must be placed. This chapter focuses on two areas of basic need – namely, the housing and food supply systems. It briefly reviews the existing situation for each and assesses their achievements.

LOW COST HOUSING PROVISION

The late 1970s witnessed several significant shifts in policy with regard to black residence in urban areas, no doubt forced upon the authorities by escalating in-migration. The first was linked to the increasing tendency for families to move into the cities, primarily to escape the ravages of the liberation struggle. As a consequence family accommodation began to dominate government housing programmes. In Salisbury between 1971 and 1977, for example, the municipality increased the share of family units as a proportion of the total provided from one-half to two-thirds. Prior to UDI the national government had long been arguing for increased availability of better housing for urban blacks as a means to bringing about greater urban stability. For the city authorities, however, such stability was not particularly welcome, nor was the anticipated consequence – namely, the emergence of a black lower middle class.

The scale of the in-migration during the 1970s, together with the growing acceptance within conventional planning circles of aided self-help housing (ASH), began to change municipal planning attitudes. Not only did ASH offer an opportunity to shift some of the costs of housing production on to the black population itself, but it also opened up the possibility to distinguish between (and divide) the black population on the basis of class by offering the inducement of eventual home ownership to those who felt they could afford it. Thus the government was forced to accept the fact that blacks should be allowed to hold freehold title to land in the urban areas. Not only was this aimed at directly satisfying middle-class aspirations, it was also an indispensable corollary to ASH programmes.

However, following research on affordability (Merrington 1981), it was argued that many of the urban poor could not realistically pay more than Z$10 per month (10–15 per cent of monthly income) and that an ultra-low-cost core (ULCC) design should be introduced. This was incorporated into the Five-Year Low-Cost Housing Programme of 1980 which comprised 61 per cent ULCC, 24.5 per cent conventional core housing and 14.5 per cent completed two and three bedroom houses. The minister responsible for housing in the new socialist government rejected the Five-Year Programme soon after coming to office. This was primarily because it was felt that ULCC housing was unsatisfactory. Certainly there had been protests in the areas where units had been built; many residents felt they were being short-changed compared to previous allocation procedures, particularly over the small garden space. But the principal objection came from the government itself which felt that ULCC somehow tarnished its image. Low-cost housing construction virtually ceased and yet squatter huts continued to be demolished. Unfortunately, the government had no feasible alternatives and in consequence housing provision fell far behind the ever-growing estimates of need. Eventually economic realism overtook public image considerations and in the Transitional Development Plan of 1983 aided self-help housing was once again made the basis for public policy, although with some modifications to the pre-1980 policies.

The aided self-help schemes have relatively high standards: a minimum plot size of 300 square metres, a detached 4-room core together with stipulated materials for walls, roof, and floor. The great majority are built for sale, only 10 per cent are for rent (usually to government employees). Given the rising costs and reduced government revenues, the average annual construction rate for the 1980s is some 10 per cent lower than that of the 1970s, and is falling rapidly behind the estimated annual target.

There were some new, socialist ideas incorporated into the plans of 1983. One was for the establishment of building brigades on the Cuban model. Three kinds of brigade were envisaged, concerned with, respectively, the upgrading of existing dwellings (particularly squatter settlements), the production of building materials, and the construction of dwelling units. The building brigades were to be organized by the local authorities which, it was hoped, would not only undertake government contracts but would be hired by self-help builders too.

Unfortunately, the houses constructed by the brigades appeared to have been more expensive than those of conventional firms, public employees, and self-help builders (Musekiwa 1989; Teedon 1990). This fact was certainly used by the World Bank and the United States Agency for International Development (USAID) – which now provide an influential 46 per cent of all housing capital – to by-pass the brigades in the schemes in which they were involved. Nevertheless, whilst the notion of building brigades seems to have comprised a good socialist initiative in the housing field, there is little evidence that the government supported the initiative with financial subsidies or in cooperative training programmes for the brigade members. Indeed, Teedon (1990) alleges that brigade development was stultified by white civil servants to whom they were anathema.

Even fewer socialist principles can be seen in the other changes introduced by the government. For example, loans for house construction were increased but only after recalculating that beneficiaries could afford to make repayments of more than one quarter of their income which was effectively beyond almost 60 per cent of the urban workforce (Butcher 1986a). In addition, although core housing was increased in size, the time periods within which dwellings on serviced plots had to be completed was drastically reduced to very short periods (Underwood 1986a). In the large Kuwadzana USAID scheme, for example, the 4-room core was to be completed within 18 months (Musekiwa 1989). The consequent pressure that these more stringent conditions have imposed on low-cost households, together with rapidly escalating costs for building material, have resulted, amongst other things, in a huge increase in lodging in order for households to acquire the money necessary to meet construction deadlines.

The question of home-ownership too has proved problematic within the context of socialist development. Contrary to what might be expected, home ownership has expanded rapidly during the 1980s led by a none-too-reluctant rush from the expanding black middle class, which included many state bureaucrats themselves. Loan opportunities have been expanded rapidly through government support, and property acquisition has become a major target for the bourgeoisie. However, the government has also sought to expand home ownership not only through new low-cost schemes but also in pre-existing projects where recommodification has been encouraged by incentives of both a positive (discounts) and negative (huge rent rises) nature.

It would seem from the evidence that government claims about 'mobilizing the people to solve the housing problem' and 'one family, one home' are simply rhetoric. So convinced is Teedon (1990) that there is little socialism in the housing programme that he dismisses such a mode of analysis and concentrates on assessing ASH programmes in more conventional ways. Essentially this involved what Simon (1989) has described as a process of replacing the 'politics of participation' by the 'politics of administration'. Even the building brigades, whilst potentially important, were envisaged merely as sources of cheap labour not as ways in which the poor could become a more complete part of socialist planning for housing provision. Little wonder, therefore, that Musekiwa (1989) could dismiss the brigades as ill-disciplined and unpopular. In particular, there has been virtually no appreciation of the problems facing those in the petty-commodity sector who have sought to meet their housing needs in less conventional ways by lodging or by squatting.

Policies towards both of these sectors has been very harsh. Threats and action against lodging, which often occurs in outbuildings, has occurred regularly since the socialist government came to power. This causes great distress and hardship both for lodgers and their hosts, particularly as lodging has steadily increased all over the high density districts since 1980. In Mbare, for example, the Joberg Lines have an estimated 10,000 people living in 681 housing units, mostly in backyard shacks (Phiri 1990). Another report by the City of Harare (1987) estimates that if the outbuildings in the low-income, high-density district of Highfield were demolished some 26,000 people or 27 per cent of its population would be made homeless.

Similar action has occurred in relation to squatters. Indeed, the response has been characterized by one minister as 'ruthless'. In 1981, some 30,000 squatters lost their homes in Chitambuhoyo in Chitungweza (see Patel 1984) and periodic purges have occurred since. The only squatter area in Harare to receive any positive government support has been the Epworth setttlement. Formerly on Mission Land and therefore relatively free from government pressures, a sprawling settlement of some 40,000 people has developed. The land has been taken over by the state and improvements are being made only to dwellings that existed in May 1983. Any subsequent construction has been quickly and ruthlessly removed.

Carole Rakodi (1989: 11) has succinctly summarized the housing programme as one in which 'an already established black bourgeoisie appears set to protect its growing property interests . . . insisting on high-quality construction for low-income residents' effectively marginalizing the poorest households.

FOOD DISTRIBUTION

The urban food distribution system in Harare has not received much in the way of direct attention from planners either before or after independence. However, information on this aspect of food marketing and distribution has begun to increase since Angela Cheater's work commenced in the late 1970s (see Cheater 1979). In some instances, such investigations have traced the movement of food from its rural origins into the city (J. Smith 1987), but primarily they have focused on the operations of urban markets and/or the informal sector (Brand 1986: Horn 1986). In contrast to even this modest expansion of research into petty commodity marketing, information on more capitalized urban retailing had been sparse, the last sustained piece of research being undertaken by Smout (1974) during UDI. Meeting food needs is an even more basic requirement than shelter, particularly for urban households for whom subsistence options seem to be limited. This means that urban expenditure patterns tend to be dominated by food and Harare is no exception to this situation. Prior to UDI urban workers were partly paid in food, and for those paid in cash some 52 per cent of their wages was spent on food compared to only 20 per cent on rent and energy (Shopo 1986). More recent surveys have indicated that low-income families still spend at least the same proportion on food, many spending up to 70 per cent (City of Harare 1987; Drakakis-Smith and Kivell 1990).

In essence, the problems facing the urban poor with respect to food are the same as for other basic needs, with the conventional supply system not meeting their requirements at a price they can afford. There are many reasons for this but most relate to the capitalistic nature of the production and supply systems (McLeod and McGee 1990). Thus, although Zimbabwe is an efficient producer of foodstuffs, much of the agricultural sector is geared towards the production of industrial and/or export crops – the recent expansion of horticultural exports is a good illustration of this trend (Smith 1987).

The corollary of this situation was a deteriorating position with regard to subsistence production in the communal lands where it was claimed that on the eve of independence only 60 per cent of food needs were being produced (Sanders and Davies 1988). Another consequence has been growing undernutrition in the rural areas, despite an expanding health care programme, particularly for families without regular employment/income (Loewenson 1988; Sanders and Davies 1988).

Although there have been few surveys to corroborate it, the situation for the urban poor is even worse. Declining real incomes have characterized the urban proletariat since 1980, whilst those in the informal sector also experience the same irregularity of income that has exacerbated undernutrition in the rural areas. Indeed, despite rural shortages, the communal areas appear to have sustained the urban poor through food transfers within kinship systems (Shopo 1986). This has been verified in recent surveys in which both low-income and middle-income households admitted to receiving food from rural areas (see Figure 4.1), usually from family sources. However, there was also a reciprocal flow of gifts, often encompassing processed foods.

The existence of this rural subsidization of urban food requirements is one indication of a problem. As with shelter this is the consequence of incomes being too low and/or too irregular to afford food supplied through the conventional outlets of the city. There are many reasons for this, ranging from the dearth of adequate income-earning opportunities to the shift of rural production away from basic foods or the changing dietary preferences of migrants to the city (see Kaynak 1981; Drakakis-Smith 1990, for some discussion of this).

One household response to the food problem within the city has been the expansion of urban subsistence production. This has been noted in other African cities (see Sanyal 1987b on Lusaka) and has been documented vividly for Harare by Mazambani (1982a). Figure 4.1 reveals the extent of urban subsistence production in home gardens from a recent survey in Harare. Little of this food finds its way into the retail system, even through the petty commodity sector; almost all is for consumption by the household itself. With the reduction of size of government housing plots over the last decade, many families have increasingly turned to illegal cultivation of the urban periphery. For many years, however, the authorities have

Figure 4.1 Non-purchased food sources in Harare

sought to disrupt such activity, even resorting to the destruction of crops.

As with housing, however, another contributory cause of the urban food problem is an inadequate distributive system. The conventional food supply system in Harare does not appear to have changed much over the last three decades (see Smout 1974). Suburbs, whether black or white, high-density or low-density, have been provided with a cluster of shops near the centre of the district, some of which deal with general or specialized food sales. Over the years, supermarkets have made their appearance, not only in the suburban retail clusters but also in the city centre, which remains an important venue for food retailing.

For many low-income housholds these conventional food retail outlets are both expensive and inconveniently located. Concern over food costs in relation to income was the principal complaint of two-thirds of all those surveyed by Drakakis-Smith and Kivell (1990). The fact that retail locations were also a problem can be seen not only by the fact that it was a common complaint, but also because both low-income and middle-income groups had devised specific responses. The latter tended to concentrate much of their shopping into the supermarkets of the CBD, either because of the greater range of commodities or because of the proximity to workplace. In contrast, low-income families, without equivalent mobility, generated a demand that was low in its fiscal and distance thresholds. This was satisfied, as in housing, by the emergence of a petty commodity sector which in Harare takes the form of a plethora of tuckshops and hawkers in the high density suburbs. The latter are almost invariably female fruit and vegetable hawkers located on tolerated and virtually fixed pitches near to the main shopping cluster. Tuckshops, in contrast, are small illegal and informal street corner stores, constructed in a variety of ways, which stock the most basic and storable food items such as bread, meal, sugar, tea, or coffee.

The existence of so many alternatives to purchasing food at conventional stores indicates the inadequacy of the formal urban food retailing system for the urban poor and yet attitudes towards and planning for this most bourgeois of urban activities indicate that the socialist government has given it virtually no attention at all. The only observable change over the last decade has been the growth and expansion of more capital intensive outlets, retailing more expensively packed and processed foods, i.e. supermarkets.

There is little evidence as yet of government attempts to socialize urban food supply and distribution, for example by the introduction of various kinds of cooperatives, in which there can presumably be much greater local involvement of personnel and commodities. Indeed, the views of the government to date on food retailing systems appear to be quite regressive in that they revolve around threats to close and demolish illegal tuckshops and prosecute hawkers, despite the obvious service they provide for low-income households. Significantly, the hawking of cooked food, so common in most other large African and Asian cities, is virtually absent from the streets of Harare (Leiman 1985).

CONCLUSIONS

After a decade of independence the achievements of the socialist government in pursuing more equitable policies of development specifically on behalf of the urban poor seem to be limited. This is not to say that some achievements have not been attained, but that judged from both a socialist and a more conventional point of view the overall results may be seen as disappointing. This is as true of the provision of other basic needs as it has been for housing and food distribution. In education, for example, although democratization of access and the removal of racial discrimination have been achieved, major problems remain to be tackled such as the shortage of qualified teachers, the persistent reliance on British syllabuses and curricula, and the fact that the state still heavily subsidizes private education (to the level of almost Z$500 per student) (Chung 1988). Ironically, expanding education is not being matched by increasing job opportunities and it is estimated that only 10 per cent of the 100,000 secondary school graduates will obtain appropriate employment (Stoneman and Cliffe 1989).

Admittedly it is difficult to extrapolate exactly what the situation might be specifically for the urban poor. However, evidence on health care provision (Loewenson 1988; Sanders and Davies 1988) indicates that the rapid growth of inflation, together with fixed minimum wages, has rapidly eroded purchasing power and so health has suffered. This presumably means that food purchases have been reduced in the face of fixed or escalated costs for other basic needs such as shelter or energy, not an unrealistic assumption given recent trends in rents and fuel prices. Furthermore, the rising urban population has produced an expansion in the petty commodity sector and as Loewenson's (1988) research clearly indicates, irregular or non-permanent work has a positive correlation with health problems, particularly undernutrition.

Observations such as these raise the question as to why the achievements of the first ten years of independence have fallen short of expectations; failing to socialize the urban masses and bring about more egalitarian access to limited social resources. To be sure, there have been considerable constraints imposed by the Lancaster House agreement and from a desire not to alienate the international capitalist interests that continue to dominate much of Zimbabwe's economy. However, in many ways these are convenient and none-too-plausible excuses. Foreign firms have usually been happy to support the sort of basic needs programmes recommended by the World Bank, seeing them as the acceptable face of capitalism and as essential for the creation of a stable urban workforce. More convincing explanations must come from an analysis of government motives tather than externally imposed constraints.

The government has made no secret of the fact that it places greater emphasis on the need for rural rather than urban development.

Indeed, it is this preference which perhaps lies at the heart of the development dilemma with regard to the urban poor. Not only is ZANU-PF ideologically committed to a form of Maoist, rural-oriented socialism, but its leaders still appear to have an innate distrust of urban dwellers whom they regard as having contributed relatively little to the liberation struggle. This distrust also extends to what is often seen as the basis for the development of urban socialism, i.e. the trade union movement, which at present is in conflict with the government and its representatives over how best to promote the interests of urban workers. Admittedly the unionized proletariat in Zimbabwe seems to comprise more of a wage elite than the raw material for socialism, but the government also displays little sympathy for, or understanding of, the non-unionized petty commodity component of the urban workforce. The consequence of all this is a virtually unopposed and expanding monopoly capital in Zimbabwe's towns and cities.

In this context, it is crucially important for the present government to begin to strengthen its programme of socialization but within a framework that recognizes the fundamental importance of the relationship between capitalism and space. Zimbabwean socialism, if this is what the government really wants to achieve, must be pushed forward on two quite different fronts – against the agrarian settler rump on the one hand, and against urban industrial capital on the other. As yet its traditional ideological inclination towards rural development has left the expansion of urban-industrial capital almost unhindered. It is because of such changes that there is an urgent need for a comprehensive framework for urban and regional development if socialism in Zimbabwe is not to lose any of the momentum it still retains. Perhaps the recent ending of the Lancaster House Agreements and the possible advent of one-party rule will bring about changes but at present this seems unlikely. More promise for improvement seems to lie in the nascent social movements that have begun to appear over the last few years.

5

SOME ISSUES IN URBAN DEVELOPMENT AND PLANNING IN TANZANIA, ZAMBIA, AND ZIMBABWE

Carole Rakodi[1]

PLANNING AND MANAGEMENT OF URBAN DEVELOPMENT

A significant proportion of national populations live in urban areas which will continue to grow rapidly even if policies aimed at retaining population in the rural areas succeed. UN estimates show that by the end of the century, if present trends continue, over half the population in Tanzania and Zimbabwe and nearly 80 per cent in Zambia will be living in urban areas. Even if such extrapolation of existing trends is unrealistic, urban areas are likely to absorb a considerable proportion of population growth (see Table 5.1).

In addition, urban areas are the location for much economic activity – which is important to national development – and act as service centres for rural regions. In order to facilitate economic development, and to maintain the health of urban populations, management systems are required which will enable cities to accommodate rapid growth, provide a viable location for economic activity, and satisfy the basic needs of their residents. African countries cannot afford to devote a disproportionate share of national resources to urban areas. Processes and patterns of urban development, therefore, need to be appropriate to the resource base of the country, to be as far as possible financially self-sustaining, and to build on existing processes for the production of housing and the economic strategies of households.

In order to devise suitable policies for the management of urban development, an understanding is needed of the market processes in land and housing in which planners seek to intervene. The urban poor have been adversely affected by deteriorating economic circumstances, but little research has been carried out on the strategies

Table 5.1 Tanzania, Zambia, and Zimbabwe: proportion of population in urban areas

	% of population in urban areas	
	1985	*2000*
Tanzania	22.3	56.2
Zambia	49.5	77.9
Zimbabwe	24.6	54.1

Source: UN (1987) *Prospects of World Urbanization 1984/5*, Department of International and Economic Affairs.

adopted by households to cope with changing conditions and the extent to which planning systems and land and housing markets hinder or facilitate such strategies. Thus any research into planning processes needs to consider the following points: first, the context in which it is operating in order to develop an understanding of the economic, political, and social processes in which attempts are made to intervene; second, the nature and form of the state, including the interests which the planning process may be serving and the nature and capacity of the state apparatus; and third, the outcome and impact of policies and their implementation.

To understand both the potential for and constraints on policy and the outcome of decisions, the urban system must be studied both 'from the top down', analysing city-wide economic and political processes and administrative structures and processes within a national context; and 'from the bottom up', assessing the opportunities and problems presented by economic, political, and administrative processes and attempts at policy implementation from the point of view of residents in households and as individuals. Such attempts at a systematic understanding of urban development are rare, partly because of the complexity of urban systems and the paucity of data, and partly because research designs rarely take into account the overall needs of urban management. In practice, research must be selective, aiming to increase understanding by designing complementary research activities, in order to build up a more complete picture of the urban system and an understanding of its dynamics and processes of change over time.

LAND MARKETS, LAND DEVELOPMENT, AND PLANNING

Little is known of the way in which the land market operates in Third World cities – especially in Africa – and assessments of the extent to which market mechanisms, together with government intervention, ensure the availability of land for urban uses are rare. Research needs to study the land development process, ways in which residents obtain access to land, land and property taxes and their role in financing urban development, land use guidance and control, and the land administration system. Formal processes for the allocation and delivery of land for urban use exist in each of the countries. The extent to which they operate efficiently in practice and the characteristics of beneficiaries need to be assessed, as does the extent to which the system facilitates the control and guidance of development. It is known that many people cannot obtain access to land through the official allocation processes, and so the means by which they do obtain access to land or housing needs to be analysed. The traditional mechanism for land use allocation and control of development is the master plan. Land use plans have been prepared for urban areas in all three countries, but systematic study of the instruments available for implementation, and the extent to which plan proposals have been implemented is rare. Explanations for the observed outcomes may include the mechanisms for land delivery, physical infrastructure provision, and the financing and structure of local government. The system of land taxation may be assessed with respect to revenue generation, financing of infrastructure provision, equity, and efficiency.

Tanzania

Dar es Salaam has grown rapidly since independence (O'Connor 1988), particularly between 1985 and 1988, leading to fragmented development, urban sprawl, and physical infrastructure deficiencies. The formal processes for the allocation and delivery of urban land are, theoretically, not a problem, as all undeveloped land is in public ownership, land has been declared to have no value, and approval has to be obtained for all transactions. The Land Ordinance applies to urban areas and recognizes rights of occupancy, provided allottees adhere to the terms and conditions of the grant. The relationship between this form of tenure and customary tenure is unclear, and there is controversy over the validity of customary tenure once land is included within the urban boundary (Tenga 1988; Fimbo 1988).

However, in practice, the process of land allocation is very bureaucratic, and the bureaucracy is unsympathic to the needs of low income people (Kalabamu n.d.; Stren 1982; Despande and Masebu 1986). Thus between 1976 and 1979, 7,703 plots were surveyed and allocated in Dar es Salaam, but 19,600 applications were outstanding, and the situation has deteriorated since then. Many of the plots have been allocated according to 'monetary, political or social influences' (Kironde 1988: 4), and the shortage is such that plots are sought with sale rather than development in mind. The plots which are allocated often remain undeveloped for reasons which are not altogether clear, but which are partly linked to the inability of infrastructure providers to service the land, partly to the absence of taxes on vacant land (Kaitilla 1987), and partly to the low level of payments for a plot, which fuels demand and provides insufficient income to finance utility provision. Although leases contain a clause permitting repossession if development does not occur within three years; this is rarely enforced. The land allocation process has failed to tackle speculation, allegedly because of the involvement of land administration officials in the illegal sale of plots (Kaitilla 1987).

At present only about 35 per cent of the houses in Dar es Salaam are on authorized plots (*Daily News*, 27 August 1988) and the situation in other settlements is similar (Kulaba 1989). Most residents obtain access to land either in low-income unplanned areas (by squatting, often on land which is unsuitable for residential development because of steep slopes or liability to flooding), in mixed-income unplanned areas (by illegal subdivision), or in peripheral areas (initially under customary tenure arrangements). The last of these is easier around Dar es Salaam, where the peri-urban area is a sparsely populated one of subsistence farming, than around, for example, Arusha, where commercialized agriculture has been more of a constraint on urban development. Once customary tenure has been obtained, an occupant can apply for regularization, either without a survey (certificate of occupancy) or with a survey (certificate of title). As areas which are not to be serviced are not surveyed, settlement depends on borehole water and forms a haphazard pattern. Disregard for land use and building regulation was exacerbated during the decentralization period of the 1970s (Kulaba 1989). Urban renewal has also commenced in Dar es Salaam, where since a period of stagnation in urban land markets ended in 1982/3, a semi-planned process of redevelopment has been occurring in Kariakoo, close to the city centre. Little is known of the exchanges in land, of the changes in land use and physical form which have occurred, or of sources of financing for redevelopment, while the implications for planning and infrastructure provision have hardly been explored.

Although there is a problem of funds to compensate traditional owners for unexhausted improvements when their land is needed for urban development, the main problem is finance for infrastructure (Siebolds and Steinberg 1982). Over the years, the revenue-raising capacity of urban councils has been reduced partly due to the emphasis placed on rural development. The funds available for infrastructure provision were particularly limited between the end of 1973 and the middle of 1978 when urban councils were abolished and the urban areas absorbed into regional administration (Mawhood 1983).

The administrative process in Tanzania gives access to considerable power. There is, therefore, a struggle to control this process, of which land allocation is a clear example. Any attempt to increase understanding of the land development process must be set within a wider understanding of class conflict and particularly the nature of the bureaucratic bourgeoisie and other urban classes (Samoff 1979). Although given the vested interests, effective reform is difficult to envisage, ways in which the land development process could be made financially more self-sufficient should be examined, in order to increase the viability of urban local authorities and enable service provision to keep pace with urban growth.

Traditional master plans have been prepared for most Tanzanian towns. There is a need for documentation and explanation of the legislative basis for plan preparation and implementation, the nature of plans prepared, the means by which implementation was expected to be achieved, and the outcome. Commentators have been concerned with the blueprint nature of the plans, in which implementation is dealt with briefly and superficially, with no estimates of resources needed or available. The plans, for example for Dar es Salaam, have been described as overambitious, unrealistic, irrelevant and inappropriate (Nnkya and Kombe 1986; Armstrong 1987). In addition to outlining strategies to deal with urban growth, the plans have recommended measures to reduce the rate of growth. These, it is suggested, signify a 'misplaced ideology' which has ignored urban development in the interests of rural areas; certainly they seem to have been singularly unsuccessful (Paddison 1988). Although plans are prepared for the local authorities by the Ministry of Lands, Natural Resources and Tourism, the councils have a habit of preparing their own investment programmes as a basis for submissions to the Ministry of Finance, Economic Planning and Development, bypassing the official land use planning system (Mtiro 1988).

Once land has been allocated for urban use, physical infrastructure provision and operation is crucial to bringing it into the development process. Aid was available for urban infrastructure during the 1970s, but the technocratic approach adopted was insufficient to solve political and institutional problems (Hayuma 1979). Shortages of funds in general and foreign exchange in particular led to major deficiencies in the operation and maintenance of urban infrastructure; for example: cesspit emptying, and domestic refuse collection and disposal. In Dar es Salaam, only 6 per cent of households were in regular receipt of cesspit emptying services in 1986/7 and 69 per cent of households, mostly dependent on pit latrine sanitation, received no service at all. Inadequate road maintenance and water supply affect not only urban residents but also productive activities (Kulaba 1989).

The local administrative system is crucial for the processes described above. The inherited system of local councils coped reasonably well with urban growth during the 1960s, despite erosion of its revenue base and autonomy (Mawhood 1983). However, the apparent failure of similar councils in rural areas, particularly following the post-Arusha emphasis on rural transformation, led to their downgrading in 1972 in favour of decentralization to the regional level. Towns were seen as exploitative and so urban areas became neglected and run down, particularly outside Dar es Salaam. Urban local authorities received less than 1 per cent of the government development budget at the beginning of the 1980s (Mtiro 1988). Even when urban decay led to their reinstatement in an only slightly modified form in 1978, the emphasis for development expenditure continued to be placed on rural development. Until 1985 they were dependent for revenue on central government; allocations were small, and even these were underused because of the lack of administrative capacity in the urban councils (Hayuma 1986; Paddison 1988). The councils are inadequately resourced, badly managed, have contradictory objectives, have a poor collection record for their main source of local revenue (a development levy on adults), and the quality of their political direction is poor (Hayuma 1986). Underutilization of existing staff because of the lack of equipment and financial resources is a greater problem than staff shortages (Kulaba 1989).

Zambia

In Zambia, a report on the land delivery system in both urban and rural areas which documents the institutional framework and procedures for land administration has recently been prepared by consultants. The cumbersome allocation process and lack of finance for infrastructure provision exclude many from the formal process of obtaining access to land. Reforms in the mid-1970s were designed to increase control over land and prices (by conversion of freehold to leasehold, a declaration that land has no value, and requirements for prior approval for transactions and the price at which land changes hands), and to improve the supply of land with secure tenure (by providing for simplified survey procedures in serviced plot areas and appropriate tenure arrangements in upgraded areas). The effects of these reforms have not been systematically evaluated. It has been suggested that together with shortages of finance they have hindered the supply of land for development, giving rise to evasion of controls. Also, local authorities are said not to be registering squatter areas as 'improvement areas' because of the obligation implied by the legislation that they must then provide utilities. Meanwhile, most low income urban residents continue to obtain access to land by renting in upgraded unauthorized areas or by squatting.

The approach to urban planning in Zambia has been similar to that of Tanzania. The development plan laboriously and expensively prepared for Lusaka has in practice been largely incidental to development in the city (Rakodi 1987a). In addition, although layout plans for smaller settlements provide neither policies to guide future decision-making nor an adequate framework for development control, resources have still to be devoted to their preparation. Legislation envisages that plans will be implemented by means of regulatory instruments to prevent unplanned development. However, the process of obtaining planning permission prior to obtaining a lease and carrying out development is slow, costly, and complicated. Evasion is widespread and there is much unauthorized development, but enforcement is limited because of the lack of staff.

The system has legal, administrative, and political weaknesses. Resistance to development control occurs both in unauthorized areas, which form the basis of political support for many politicians, and where local authorities are seen not to observe the regulations themselves (Rakodi 1988a). The procedures for plan preparation are too rigidly stipulated and plans are too inflexible to accommodate rapid urban change; they are unrelated to economic planning and budgeting at the national level and so investment for realization of the plan proposals is rarely forthcoming, while administration of the development control system absorbs a disproportionate amount of the planning resources available and is clearly largely ineffective, although there have been no published assessments of implementation methods and achievements. The planning framework is, therefore, based on a largely negative system designed to prevent unplanned development, but without the necessary powers to promote or guide desirable development.

A major obstacle to improvements is the lack of political and public support for planning. Planning is viewed as unnecessary interference in private property rights. Land has no value and so benefits do not flow to individual owners from enforcing planning standards, and the perceived failure of the system to deal with urban problems further reduces its legitimacy. Although periodic bargaining yields some benefits for the poor, the absence of political pressure for the system to have a more extensive role in basic needs satisfaction or redistribution allows its use by the elite in maintaining its own standard of living. Planning's poor performance thus has to be explained in terms of the conflicting interests of the wider political economy, as well as in terms of resource constraints and inappropriate planning approaches. Despite this experience, a review of the Lusaka plan is being considered (Vagale 1987). Although this takes on board developments in approaches to planning in the 1970s and recommends the preparation of a broad strategy and more detailed shorter-term programmes and projects, it does not really come to terms with the lack of efficacy of the planning process.

Local administration was reformed in 1981 (Rakodi 1988b). Local authorities were, theoretically, given more autonomy and a greater degree of financial self-sufficiency, but in practice strict central government control is maintained and responsibilities are not matched by revenue raising powers (Chitoshi 1984). There has been neither a transfer of functions from central to local government, nor a different structure of accountability for the local staff of central ministries because of resistance at the centre, while local government staff resent the erosion of local democracy which they see as having occurred (Lungu 1986). The relationship between the administrative and political structure is crucial in understanding the poor performance of local government. Prior to 1981, standing for election as a councillor was attractive to professionals and businessmen, so that councillors brought some relevant expertise to their council work and were not dependent on the income from their jobs as elected representatives. The new generation of councillors, are ward chairmen (i.e. local party activists and officials), depend on their allowances for their primary source of income, are attracted by the status of elected office and see local authority work as a stepping stone to a political career. Officers see them as less expert and more likely to give priority to their political careers than to local affairs. The interests served by pre- and post-1981 councillors are rarely compared.

The financial position of local authorities is crucial for service provision and infrastructure operation and maintenance. Although some major infrastructure is provided and maintained by central agencies (highways, electricity), that for which local authorities are responsible is poorly maintained and provision lags behind urban growth (see Cheatle 1986; also Blankhart 1986, and Mwali 1988 on public transport). Thus all urban areas face infrastructure problems and severe resource constraints, as well as limits on their autonomy. In addition, political adjustments are needed to come to terms with the new system of representation. Although currently there are attempts to strengthen the planning and policy formulation capacity of the Ministry of Decentralization, there appears to be little recognition either of the problems faced by urban local authorities or that solutions may exist. The potential contradictions inherent in central–local state relations have led to the erosion of local autonomy and enhanced financial and administrative control by central government. The lack of local authority power to raise revenue and make decisions, and the confused legislative basis for the performance of local government functions, has thus prevented the development of local administrative capacity (Rakodi 1988b).

Zimbabwe

In Zimbabwe, concern over the cost of low density urbanization, the need to provide serviced land for low and middle income housing, the inherited segregated urban structure, and the financial basis for local government and development is revealing problems related to the ownership and use of land and planning for development. Yet the nature of the land market and procedures for making land available for development were not, until recently, seen as problematic. The land allocation and development processes have not been documented, trends in land prices are not known, and there has been little analysis of the land and property taxation system in relation to the financing of infrastructure. At the time of independence, utilities in the main towns were still well maintained, and despite the loss of qualified staff, the operational and financial capacity for maintenance has been preserved. Shortages of foreign exchange have had most impact on the public transport system, giving rise to difficulties of maintaining vehicles on the road in order to meet demand (Situma 1987). Following reconstruction of war-damaged facilities in some of the smaller towns, physical infrastructure is also operational there.

The formal land allocation system based on private tenure continues to operate. In the high income (former European) parts of the urban settlements, stagnation of population and construction during the later 1970s and emigration on a considerable scale immediately before and after independence meant that until recently there has been little pressure on the system. In Harare and Bulawayo, the scale and nature of problems is similar, while there has been a deliberate policy to develop lower order settlements (Gasper 1988). The issue is whether the economic and administrative systems will facilitate the latter policy. Problems with the underlying assumptions on which such a strategy is based have resulted in only limited growth in the designated centres. The contribution of a physical planning system based on layout preparation and development control has been limited to date.

In the low income areas of the cities, land allocation was a public sector activity until in-migration during the war gave rise to squatter settlement. Although the official processes of allocating and servicing land for low-income housing could not keep pace with urban growth then, and have not been able to do so since independence, squatting has been strictly controlled and most squatter areas have been demolished and their occupants resettled. Epworth, a settlement on former mission land just outside the city, is the sole exception in Harare (Butcher 1986). To date, this role of government has been upheld by politicians and squatters have been unable to resist. As a result, continued in-migration has resulted in multi-occupation in the existing housing stock rather than unauthorized construction of new houses. Effective control of unauthorized development has tended to conceal problems of land and housing supply for low- and middle-income people, which manifest themselves in overcrowding of the existing housing stock and extensive illegal lodging (Rakodi forthcoming). New development for low income housing to date has occurred on land in public ownership, or agricultural land which is offered to the government, giving rise to a rather *ad hoc* pattern of urban development. Under the Lancaster House Agreement, until 1990 any compulsory purchase of land has to be compensated in foreign currency, inhibiting the scope for land acquisition by the government in urban as well as rural areas.

The planning system is, typically, based on the preparation of master and local plans, following complex procedures and subject to precise but fairly meaningless requirements for public participation laid down in the Regional, Town and Country Planning Act of 1976. By the mid-1980s, only one master plan and twelve local plans had been approved, many local authorities still continued to rely on town planning schemes prepared under the 1945 Planning Act, and several master plans had been in preparation for years (Underwood 1986b). Although the Harare development plan is currently under review, there has been no published evaluation of the existing plans for Harare and Chitungwiza which it is to replace, and no assessment of the extent to which the proposals contained in those plans have been successfully linked to decision-making and implementation mechanisms. That there is considerable practical implementation capacity, in terms of the physical development of planned low income housing areas on public land, is demonstrated by the construction of nearly 30,000 houses and associated facilities in the new town of Chitungwiza within the last decade or so, comprising nearly a fifth of the total housing stock of the conurbation. The new Combination Master Plan is being prepared in-house rather than by foreign consultants, which must to some extent improve its chances of implementation.

Urban local government retains its pre-independence structure, with elected local councillors and a considerable degree of continuity in practice, despite the changeover in political representatives and senior officers. Hodder-Williams (1982) describes the effects of the election of predominantly Black councillors in Marandellas, a small town 40 miles south-east of Harare, in 1980. The perceptions of their political role held by the new councillors differed considerably from their predecessors, with implications for their concept of the functions of local government and political and administrative practice. The political imperatives which governed their actions, coupled with inexperience, led rapidly to crises over spending and over the form of local government and the values it represented. The study was of an urban administrative system in flux and has not been followed up, in Marandellas or elsewhere, by comparable research.

The urban local government system is clearly strongly established and efficient, especially in the relatively autonomous larger settlements. Councils have a wide range of functions backed by significant budgets and considerable capacity for local revenue generation. Since independence, however, growing deficits have absorbed accumulated surpluses, and both the balance between central government financial control and local autonomy and the land and property taxation system need to be reconsidered (Wekwete 1989). The main account of local government (Jordan 1984) contents itself with technical description and limited evaluation of administrative efficiency, and deals neither with the political nor inter-institutional relationships and ideology which influence decisions made by local authorities.

HOUSING MARKETS

Most research has concentrated on unauthorized areas and the outcome of policies to improve housing conditions for the poor, some of which have, in practice, benefited middle-income groups. A more systematic understanding of urban housing markets is needed, including the private sector production of housing, public sector roles in house production, and constraints both on the supply of housing and on access to housing by all income groups. Policies are needed which will enable all groups in the population to satisfy their housing needs and prevent hijacking of provisions intended to benefit the low income groups by those further up the income distribution whose needs have not been met.

Tanzania

In Tanzania recent efforts in housing provision have concentrated on the provision of plots in sites and services schemes and upgrading of squatter areas, with World Bank assistance, and research has concentrated on the outcomes of these programmes. Despite the low priority given to the urban sector, especially following the abandonment of urban local government in 1972, some policies addressed the housing needs of urban residents. Policies to keep wages from rising included the nationalization of rental buildings, the imposition of rent controls, the provision of sites and services, and upgrading of unauthorized areas (Kulaba 1981; Siebolds and Steinberg 1982). The first of these policies yielded some rent revenue for the Registrar of Buildings, but inhibited private investment in the construction of formal sector housing for rent until it was waived in 1987. Meanwhile, extensive letting of rooms in unauthorized areas, mainly by owner-occupiers and not solely to meet the housing demand of the poorest, continued (Magembe and Rodell 1983).

Before 1972, housing provision had concentrated on the construction of flats for rent by the National Housing Corporation, but between 1962 and 1980, only 13,366 units were built and the average rents exceeded the capacity of low-income households to pay (Mghweno 1984; Kulaba 1981). The first phase of the World Bank funded project (1974–7) including the servicing of 10,600 plots in Dar es Salaam, Mwanza, and Mbeya, upgrading in Dar es Salaam and Mwanza affecting 8,800 dwellings, and financing for a Tanzania Housing Bank (THB) soft loans programme for low-income households. In addition, a review of building regulations was required. However, the process of obtaining a plot and then a loan was long and bureaucratic, required the payment of fees, proof of regular income, and savings of six times the average monthly income of low wage employees. Although the plots were criticized for their small size (12 × 24 m), those provided met the needs of middle- and upper-income groups rather than the low-income households for which they were intended (Siebolds and Steinberg 1982), and allottees were able to invest significant funds of their own in construction over and above the THB loans (Magembe and Rodell 1983).

Phase II (1977–80) extended the programme to Tanga, Tabora, Iringa, and Morogoro, in addition to Dar es Salaam, including serviced plot schemes and upgrading, both on an ambitious scale. Thus it was hoped to benefit 26 per cent of Tanzania's urban population, upgrade 40 per cent of existing squatter settlements, and satisfy the demand for 75 per cent of residential building land needed during the project period. The project continued the provision of THB loans for house construction and improvement and included further institutional development and training. Despite the stated preference for low-income people, and a points system for allocation of plots which favoured them, the downpayment requirements and cost favoured middle-income groups (Siebolds and Steinberg 1982). Mghweno (1984) estimated that at least half of all urban households had a maximum of TSh150/month available for housing and that this was roughly equal to the plot charges alone. In addition, implementation was delayed by general economic difficulties and shortages of personnel and funds, so that by 1981 less than 30 per cent of the plots were ready for occupation (Mghweno 1984). Several researchers have evaluated the serviced plot programme, and all come to similar conclusions: that the programme exceeded the capacity of the target low-income group to pay, and that implementation was slow and further hindered by complex allocation procedures and shortages of building materials (Kulaba 1981; Stren 1982; Magembe and Rodell 1983; Fair 1984; Kamulali 1985; Materu 1985; Sheriff 1985; Schmetzer 1987; Campbell 1988). The distribution of benefits from downgrading has been more widespread and has not been regressive, but standards of infrastructure provided were low and operational deficiencies are widespread, not least because of the centralized planning and design of the programme and failure to involve local authorities and residents (Magembe 1985).

Although a third phase was intended, the World Bank withdrew from funding apparently because the serviced plot areas on which very substantial houses were being built, except in the least popular areas, were not reaching the target group for reasons of both cost and corruption in the allocation process. The programme may also have been a casualty of the protracted arguments between Tanzania and the IMF over appropriate adjustment measures. Currently, a small government-funded programme is being implemented by the Ministry of Lands in Dar es Salaam. Demand for the plots comes largely from the middle-income groups, as plot sizes have been increased, and individual water and electricity supply is provided (although 90 per cent have pit latrine or septic tank sanitation). Although THB loans are available, in practice substantial houses are being constructed largely using the resources of participants themselves. Leakage up the income distribution is encouraged by the shortage of appropriate houses in the private sector and the absence of restrictions on resale. The size of the programme is said to be constrained by the finance available, the urban planning system, and shortages of staff and building materials.

The Tanzanian experience illustrates both the difficulty experienced if housing programmes are divorced from the rest of the urban management system and insufficient attention paid to necessary inputs, and the determination of their outcomes by the bureaucratic and class structure in which they are situated. One of the drawbacks of reliance on aid, Hayuma (1979) suggests, is its *ad hoc* contribution to urban development in the absence of a policy framework. The serviced plot programme has benefited the dominant classes and private capital directly, via both the award of contracts for construction and infrastructure installation and the allocation of plots, and indirectly by the increase in land values and rents which has occurred. The extent to which the nature of these benefits can be attributed to World Bank domination of planning and implementation (Campbell 1988), and the extent to which it would have occurred anyway in the Tanzanian situation, is open to argument. Campbell blames the World Bank for cutting and running when large cost overruns resulted from delays in implementation and national economic problems, in the process handing over administrative responsibility to local institutions which have been incapable of managing the areas. Cost recovery is poor in both serviced plot and upgrading schemes – unsurprisingly, as local government lacks the capacity to maintain infrastructure (Kamulali 1985; Magembe 1985; Kulaba 1985, 1989). However, as we have seen, the lack of commitment to institutional development of local government is a deep-rooted problem based in Tanzanian ideology and policy, and cannot be solely blamed on the World Bank in this case. Other than the research on the outcome of the World Bank funded programmes there have been few studies of other aspects of house production in Tanzania. A more thorough review of available material is needed to summarize the state of knowledge and reveal gaps in understanding of housing markets.

Zambia

In the first decade after independence in Zambia, housing policies emphasized the construction of complete houses for rent to low-income household and Civil Servants, and the provision of serviced plots. However, the rate of urban growth was unprecedented and, despite relative prosperity, the policies adopted and limited administrative capacity prevented the urban areas, even in this relatively urbanized country, absorbing their rapidly growing populations. As a result, by the beginning of the 1970s, a third or more of the urban population had accommodated itself in unauthorized areas, including both settlement with permission of the landowner on disused farmland on the urban periphery and squatting on land in public ownership. The evident failure of housing programmes, particularly the construction of complete houses, led to a reconsideration of policy in the Second National Development Plan. Emphasis was increasingly placed on the provision of serviced plots and, for the first time, on upgrading unauthorized areas. Implementation of this strategy in Lusaka, with the aid of a World Bank loan, took place between 1974 and 1981 and has been extensively documented.

Household sample surveys, particularly those carried out during the evaluation of the project in Lusaka, gave a reasonably good picture of socio-economic characteristics, housing conditions, and access to infrastructure of residents of upgraded and serviced plot areas in the 1970s (Bamberger *et al.* 1982; Sanyal 1987a; Rakodi 1986, 1987b, 1988c). Two standards of serviced plots were provided: most with individual taps, road access to each plot, and waterborne sanitation, and a minority with shared taps and pit latrine sanitation. The former in particular met part of the demand for medium-cost housing and, while accounting for a large proportion of the total cost of the project, accommodated a much smaller proportion of total beneficiaries. Many more households benefited from the installation of physical and social infrastructure in the major squatter settlements around the city. The adoption of relatively generous standards, especially for road reserves, gave rise to a considerable need for resettlement of households whose houses were affected by the programme. This need was met by developing resettlement areas adjacent to the upgraded areas and issuing loans for house construction, although many of the poorer, more vulnerable and smaller households chose not to avail themselves of the opportunity to build new and better houses. Administration of the areas has now been institutionalized within the urban district council, although cost recovery and infrastructure operation and maintenance are continuing problems.

Although official housing policies have not changed, implementation in Lusaka has been confined to small-scale construction for sale and the upgrading of further unauthorized areas – one with German financial and technical assistance. Similar policies are being pursued in other settlements, but have been almost entirely dependent on the availability of aid funds – from the European Development Fund, for example – and details are not available. The studies in Lusaka dating from the 1970s have been neither updated, except with reference to very selective themes, nor set in the context of wider housing market studies. Schlyter (1987a), for example, has examined the development of market relations in an upgraded squatter area, and found that rental submarkets are important even in housing areas which are still largely owner-occupied and self-built. She has also documented the disadvantaged position of households headed by women in the housing market (Schlyter 1988).

Urban growth seems to have been accommodated by both new squatter development and increased densities in upgraded areas, thus increasing renting and overcrowding. Shortages of low-cost houses have given rise to sub-letting of housing provided by employers and high rents in the private sector. Recent research on towns other than Lusaka is limited, though see, for example, Kasongo (1987) on Kitwe. Responsibility for housing was dropped from the reconstituted Ministry of Decentralization and the absence of a national housing policy is attributed to the fragmented institutional responsibility, and also to failure to recognize housing construction as an activity potentially capable of contributing significantly to economic growth.

Zimbabwe

A review of published and easily accessible secondary material on the production and consumption of housing in Harare identifies problems in various segments of the housing market and points to a need for further information to guide policy-making in these areas (Rakodi 1989, 1990; Rakodi and Mutizwa-Mangiza 1989).

Since independence racial barriers to residence in *low density areas* have disappeared, giving rise to considerable in-movement by Black Zimbabweans, particularly in the early 1980s (Harvey 1987a) when excess supply depressed house prices. However, failure of wages to keep pace with prices, together with a tightening of the market leading to rapidly increasing house prices, has resulted in demands for public transport services, subletting of rooms and servants' quarters, and increased difficulty – even for relatively high-income households – in obtaining access to housing. For these and other reasons, densification is being considered. There is a need to study the processes of change in typical low-density areas and to explore the social and technical feasibility of densification.

With regard to *middle/high-cost rental housing*, rent control was introduced in 1982 and has inhibited new construction for rent and led to the sale of flats and apartment blocks. This process of exchange needs to be documented, showing the socio-economic characteristics of landlords, owner-occupiers and tenants, and assessing satisfaction with dwellings occupied.

Middle-income households are currently accommodated as sub-tenants in low density areas, tenants in apartment blocks, and owners or tenants in municipal housing or serviced plot areas. In order to reduce the shortfall in the supply of housing they can afford, the Ministry of Public Construction and National Housing intends to initiate large-scale construction of walk-up flats. The Ministry proposes that 60 per cent of all new residential development should be in flats in order to reduce urban sprawl. It appears to think that the long-term costs of sprawl exceed the short-term cost of multi-storey construction, and that full cost recovery can be achieved for middle-income groups while local authority assistance will only be required for those earning less than Z$500/month. The current housing situation, socio-economic characteristics, and the housing needs and aspirations of middle-income households need to be analysed, with a view to assessing the feasibility of this solution and recommending alternatives if appropriate.

Although it has always been acknowledged that some *low-income housing* needs will be met by subletting (registered lodgers), municipal house provision and serviced plot programmes were also intended to combat this problem. Today, there are nearly 110,000 such units in Harare and Chitungwiza, accommodating about 60 per cent of low-income households, as owners or main tenants. In practice, problems of affordability and shortfalls in supply (there were 46,000 households on the waiting list in Harare in 1988) have given rise to extensive illegal subletting and overcrowding, the extent of which needs to be assessed. The socio-economic characteristics and housing needs and aspirations of lodger households need to be ascertained, with particular attention to the needs and difficulties of households headed by women (Schlyter 1987b), to provide a guide for policy.

Most former municipal rental housing was sold to sitting tenants in the early 1980s, while serviced plot schemes were designed for owner-occupation. Restrictions on individual property ownership and resale only apply while plot occupants are repaying the initial cost of purchase and construction. Even at this stage, the restrictions may be evaded, and once houses are owned outright these procedures do not apply. The extent of resale and the resulting patterns of ownership and tenancy need to be analysed in order to assess who has benefited from these housing policies (see also Teedon and Drakakis-Smith 1987; Teedon 1990).

HOUSEHOLD SURVIVAL STRATEGIES

There is only patchy knowledge of the coping strategies adopted by urban households, based on occasional household sample surveys and government household budget surveys, both of which provide a static and oversimplified picture. Research is needed which reveals the dynamics of household strategies over time, analyses the contribution of both men and women in the household to income earning and reproduction, and examines the means by which households obtain access to land, housing, and services.

The recent UNICEF report, *Adjustment with a Human Face* (Cornia *et al.* 1988), is concerned with the distributional and welfare impacts of structural adjustment lending (SAL) programmes. It shows that one of the groups adversely affected by the macroeconomic policies which generally form the basis of such programmes is the urban poor, who suffer as a result of increased food prices, wage freezes, and cutbacks in expenditure on basic services. The report recommends various ways of ameliorating the adverse impact of SAL policies, but recognizes the limited research on which its conclusions and recommendations are based. This is especially true for urban areas, where the coping strategies devised by households during the worsening economic circumstances of the 1980s have been relatively little documented.

Welfare depends on the ability of households to satisfy their needs with respect to income generation and/or production for self-provisioning, access to basic infrastructure, services, and housing (land and shelter). The strategies adopted by households change over time in response to changing economic circumstances and policies, in an attempt either to survive or take advantage of opportunities to increase incomes, wealth, and security in the longer term. Relatively little research has been carried out on the livelihoods of low-income households and the extent to which planning systems and land and housing markets hinder or facilitate their coping strategies. Households are likely to have diversified sources of income, including activities engaged in by women and even children. It is expected that migration behaviour and the relationship between urban households and rural kin, especially rights to rural land and flows of money and goods between urban and rural areas, may have changed. Studies need to include both typical households and groups which are disadvantaged, such as households headed by women; to seek information from both male and female adult household members; and to look at changes over time.

Tanzania

Some information on access to land and housing is available from the results of the Ardhi Institute surveys (Kulaba 1989), but these do not show how households and their members have adopted to the economic changes of the 1980s, or the survival strategies of the poorest, including relations with rural areas, migration patterns, and economic activities. The impact of the crisis on urban households has been essentially twofold. Inflation has exceeded the growth in incomes, so that by 1980 the minimum wage was 20 per cent of its 1974 value and was already then insufficient to provide a month's supply of food for an individual. By 1984, it was estimated that families depending largely on one member earning the minimum wage lived on no more than three-quarters of their calorie needs. Real wages continued to fall as a result of wage freezes in the 1980s, although total employment apparently did not fall (International Labour Office 1988). Households, therefore, have been forced to diversify their income-earning opportunities, and many have experienced a decline in their standard of living, although the proportion of households which have maintained their standard of living and the proportion who have suffered cuts is not known (Tibaijuka *et al.* 1988). Little is known about the growth of the parallel economy and the extent to which it has cushioned falling real wages (International Labour Office 1988; Biermann 1988; Kulaba 1989). In addition, since the late 1970s the government has been forced to cut back expenditure on social services, in both absolute and relative terms. These cuts in services such as health care and water supply have given rise to increased morbidity, stagnating or increasing infant and child death rates, widespread malnutrition because of decreasing incomes, increased pressure on women to engage in income earning activities, and an apparent increase in child labour (Tibaijuka *et al.* 1988). Shifting responsibility for primary education from central to local government and inadequate funding have resulted in declining primary school enrolment as a percentage of the eligible age group (International Labour Office 1988).

A UNICEF/SIDA funded study of the socio-economic effects of adjustment is planned for three to five years from 1989, in which data will be collected on socio-economic indicators (status indicators such as infant mortality, process indicators such as the incidence of malnutrition, and input indicators such as household food shortages) in order to increase understanding of the processes and their causes, in five rural and two urban areas (Tibaijuka *et al.* 1988).

Food supply is clearly an essential element in household survival, and urban supply depends on national agricultural production and policy and on the attempts of urban families at self-provisioning, via both rural links and urban production (Bryceson 1984, 1985a, 1987). Until 1984 maize meal was subsidized and, as it was often distributed at the workplace, especially benefited wage workers and their families. In that year, the subsidy was abolished, but the shock was largely cushioned by a simultaneous increase in wages (International Labour Office 1988). The food deficit which affected towns and cities in the 1980s was not the first to be experienced, and has given rise to the use of parallel sources of food. Rural sources include the unofficial grain market, supplies from relatives, and cultivation of a farm in a household's home area. Cultivation within or on the periphery of urban areas is also widespread, although the assumed health hazard is apparently felt by some to outweigh its crucial contribution to family welfare (Kulaba 1989). Although the colonial concept of a family wage has been overtaken and current wages are insufficient to support a family, the unacceptability of engagement in wage labour by women has meant that their economic activities are confined to extensions of their domestic roles – including subsistence food production – which makes a significant contribution to the livelihood of urban households, particularly in secondary cities (Bryceson 1985b).

Around 80 per cent of urban residents depend on charcoal or mixed sources of energy for cooking and lighting. Increases in the cost of kerosene led to a return to the use of wood and charcoal, driving up the prices of these commodities in turn. Of the electricity produced in the country in 1979, 90 per cent was consumed in the seven largest urban centres (Nilsson 1986). However, relatively few of the poor have access to it.

Zambia

In Zambia formal sector wage employment grew until the late 1970s, but has since declined and it is estimated that up to a quarter of the total labour force and between 11 and 16 per cent of the men available for work were already self-employed at that date. However, many of the typical informal sector activities, such as tinsmithing, furniture making and manufacturing, were, it was considered, essentially saturated by then, although the retailing sector was still increasing in size and there was increasing evidence that a black market in price-controlled essential commodities was growing. The government's attempts to regulate market-based trade were designed to manage the economy to ensure that basic foods and convenience goods were available, to clamp down on black market activities, and to control the traders. Its reorganization of markets and establishment of cooperatives was justified in terms of its desire to protect local (increasingly female) traders from competition from traders from other countries and to foster self-reliance. Price control was a result of its desire to ensure equal access to food and basic goods. Despite the loss of autonomy of markets, there still appeared to be the capacity to generate additional jobs in the late 1970s (Scott 1985), although the position of women traders was difficult economically and socially (Schuster 1982).

In addition, many households were able to supplement their incomes by cultivation of plots in and around the urban area (Jaeger and Huckaby 1986; Rakodi 1985, 1988d). Supplies of energy are also essential to household survival. Traditionally, low-income households in Zambian urban areas have used charcoal for cooking and heating, giving rise to depletion of timber reserves and illegal felling in forest reserves easily accessible to the urban areas. A twenty-year programme to extend electricity supplies to sixteen low-income serviced plot and upgraded areas started in 1987 (Kalapula 1987). From the point of view of government, this programme is feasible because electricity generated from hydro-electric schemes is abundant, but connection and consumption costs may be beyond the reach of many low-income households. There have also been attempts to develop and market improved charcoal stoves.

In 1979, a survey found that about a quarter of the dwellings in peri-urban settlements accommodated home-based enterprises, which raised the incomes of these households 11 per cent above those of households without such home-based enterprises by exploiting their location with respect to local markets and the labour available in the households (Strassman 1987). Since then, the supply of food and goods has worsened, real wages have decreased further, and parallel trading channels persist. A particular difficulty is the retrenchment of employment in the mining sector (Burdette 1988), which is already occurring – especially in Kabwe – both in order to increase efficiency and in response to the exhaustion of existing mines, and is likely to persist into the next century. Liberalization measures adopted in the 1980s have led to a fall in formal sector employment, declining real wages, increased unemployment, and a decline in the quality of food bought by the urban poor (Colclough 1988). It is likely that the demand created for informal sector goods and services from wages generated in the formal sector has stagnated and declined, giving rise to a recessionary spiral within the sector, although there is evidence of some new activities such as manual stone crushing. Recent studies concentrate on evaluating policies to encourage and facilitate the development of informal sector activities in Lusaka, especially the Economic Promotion Unit which provides loans for small-scale enterprises in Kalingalinga (Maembe and Tomecko 1987), and not on analysing the fortunes of the small-scale sector as a whole.

Zimbabwe

Wage employment in the six largest urban areas grew by about 18 per cent between 1980 and 1988 (Zimbabwe 1988), but this was insufficient either to absorb the growing labour force or to offset the loss of rural wage employment. The government's priority in dealing with economic difficulties since 1982 has been to restore the external economic balance and maintain the economy of the country intact, as well as to honour its pre-independence pledges to the peasantry. Industrial unrest has been quelled (Moyo 1988; Wood 1988), but little progress has been made with industrial democracy (Sachikonye 1986), and the structure of the economy has changed relatively little, although the extent of government control has increased (Stoneman 1988). For example, firms cannot hire and fire without government permission so that, although stability of employment for those in wage employment has increased, few jobs are available for new entrants to the labour force (Kadhani 1986; Stoneman 1988).

As far as it is possible to tell, about three-quarters of the urban labour force is in wage employment, but precise figures are not available. There have been some recent studies of informal sector activities and food supply and marketing in Harare (Brand 1986; Sijaona 1987; Drakakis-Smith and Kivell 1990), but no other studies of urban livelihoods. These activities are regulated by the public authorities who restrict entry, although unlicensed activities persist (Brand 1986). In order to survive on incomes which have not increased in real terms since the 1970s, despite government's attempts to increase the minimum wage level (Bratton and Burgess 1987), low-income households maintain rural links (Potts 1987), cultivate land in and around the urban area (Mazambani 1982a), and collect fuel from woodland around Harare (Mazambani 1980; 1982b), although the standard of social services continues to be high in urban areas.

CONCLUSION

The severity of problems arising from rapid urban growth is greater and the capacity of government to perform necessary management tasks is less in Tanzania and Zambia compared with Zimbabwe. In the first two countries much of the demand for land and housing is satisfied by market forces, while the limited management capacity of central and especially local government results in failure to regulate urban development, inadequate infrastructure provision, and arrangements of land uses which may benefit individuals but give rise to costs for urban society as a whole. The infrastructure which has been provided and the services which are operated, despite the attempts of the 1970s and early 1980s (largely funded by the World Bank) to extend provision to lower income residents, tends to benefit middle- and upper-income groups disproportionately, thus perpetuating the older race/class distinctions.

The rural bias of policy and the abolition of urban local government in Tanzania in the 1970s is reflected in the greater shortfall in new infrastructure construction and more serious deterioration of existing infrastructure in that country's urban settlements than in those of Zambia, although the latter's are not far behind. The rural bias of policy, decline in real wage incomes, and comprehensive failure of urban management in Tanzania seems to have given rise to a greater variety of household survival strategies and a proliferation of non-capitalist economic activity. The apparent vitality of small-scale economic activity seems to have increased following the economic liberalization policies instituted by Tanzania in the mid-1980s. The sector appears to be less well-developed in Zambia, but the data are available neither to confirm this apparent difference nor to explain it. Differences in the extent and complexity of small-scale economic activities may, in addition, be related to city size, references to economic stagnation being more common with respect to towns than larger cities in both countries.

In both Tanzania and Zambia, the central government's fear of politically strong urban local government and its desire to maintain control over revenue generation and expenditure, both understandable concerns to some extent, have blinded it to the necessity for and advantages of self-reliant local administration. As a result, the needs of urban areas are seen primarily as a drain on central revenue and both the productive benefits of having adequate and reliable utilities and services and the capacity of urban areas to generate revenue have been underestimated. In neither Tanzania nor Zambia does any serious thought appear to have been given to the nature of urban local authorities when reorganization occurred and despite the lip service paid to their need for greater autonomy and financial self-sufficiency, central government control is maintained. Shortages of resources, especially foreign exchange, and ceilings imposed by central government on expenditure and wages are reflected both in shortages of equipment and spare parts, and in the diversion of much of the energy of local authority employees to supplementing their basic wages through other income-generating or productive activities, both legal and illegal. The deficiencies are manifest in all the urban services: sanitation and refuse disposal, water supply, education and health, energy supply, and roads and public transport.

The obvious lack of capacity of local administration with respect to its developmental and service delivery roles reduces its ability to perform regulatory tasks, as its weakness is mirrored by the lack of resident support and respect for its role in enforcing land use, health, building, and traffic regulations. Strong local government may, if it does not rest on a base of popular support, give rise to local opposition; the same may apply where local government is weak and unable to complement its task of regulating the exercise of private property rights with the delivery of adequate services. Legitimacy crises, as is obvious from studies of the central state in Africa, characterize both strong authoritarian government and weak or soft states.

Zimbabwe is very different. Urban land development is still a tightly controlled process, enforced by strong local authorities with an adequate legal basis for their actions. Infrastructure standards continue to be high and only land which can be serviced is released for development. Local authorities not only have considerable administrative, technical, and financial capacity but also are able to exercise their regulatory functions, backed up by enforcement powers and public attitudes which do not question these powers.

Nevertheless, and despite the considerable inherited and maintained capacity of Zimbabwe's local authorities to manage urban development, the system is under stresses similar to those experienced in Tanzania and Zambia. Rapid rural–urban migration is also occurring in Zimbabwe, despite moves to discourage it, while shortages of financial resources, especially foreign exchange, result in an inability to purchase and maintain equipment and vehicles. In a largely administratively determined land and housing market, backed by strong enforcement powers, the market responds to new low-income housing demand not by the growth of unauthorized housing areas but by multiple occupany within the existing stock. The strains this imposes on the land allocation, house construction, infrastructure provision, and service delivery systems can currently be accommodated and do not manifest themselves in breakdowns of these systems; but unless government responds in a realistic way to urban population growth and declining real incomes, some of the symptoms of breakdown so widespread in Tanzania and Zambia may start to appear in Zimbabwe. Perhaps there will be popular protest against inadequate public transport services, more widespread evasion of the licensing controls over street selling and other informal sector activities, or a resurgence of squatting.

NOTE

1 This chapter builds on earlier research by the author in Zambia and Zimbabwe. Additional material was collected during a trip to the three countries in 1988, for which funds were received from the Economic and Social Research Committee of the Overseas Development Administration, London. The assistance of colleagues in government and academic institutions in all three countries is gratefully acknowledged, but the views expressed here are mine. A shorter version of this chapter has been published, with permission, in the Review of Rural and Urban Planning in Southern and Eastern Africa, 1990, 78–110.

6

RURAL DEVELOPMENT POLICY, REGIONAL DEVELOPMENT PLANNING, AND DECENTRALIZATION IN BOTSWANA

Paul van Hoof

INTRODUCTION

In many developing countries, regional development has been introduced in order to improve the relation between planning and rural development. The government of Botswana has adopted such a regional planning approach for rural development since the mid-1970s.

This chapter describes the character of regional planning for rural development in Botswana. Within this an analysis is made of the way in which elements of the national planning environment, the regional differentiation of natural and human resources, and the character and degree of the decentralization of decision-making power have influenced the content and character of regional development planning in Botswana. First, attention will be given to the relation between rural development, regional development planning, and decentralization in general.

RURAL DEVELOPMENT, REGIONAL DEVELOPMENT PLANNING, AND DECENTRALIZATION

For several decades, governments in most developing countries have attempted to stimulate and guide rural development through policies and planning. Over the years, the views on rural development and the content of rural development policies have changed from 'rather simple and straight forward growth orientation to an increasing awareness of the complexity of the issue of development' (Sterkenburg

1987: 17). Sterkenburg defines rural development as a process of change in the rural areas leading to better living conditions and a greater security of existence for the population. This process comprises:

1 The growth of production and productivity, and the diversification of production activities within the agricultural sector.
2 The increase in complexity and linkages in the rural economy as a result of the expansion of non-agricultural production activities, and rural industries in particular.
3 The expansion and amelioration of agro-support and community services.
4 The improvement of environmental conservation as a form of preserving natural resources, which is essential to sustain the process over a longer period of time.

(Sterkenburg 1987: 20)

Government policies concerning rural development have evolved as well. In many countries, such as Sri Lanka, Indonesia, Kenya or Botswana, this multifaceted character of rural development is recognized and used as a premise for rural development policies with a more comprehensive and integrative character, addressing several or all of the above-mentioned objectives at the same time. However, many differences exist concerning the ways and means to achieve these objectives, stressing more or less the importance of production increases, the strengthening of social services, the changes of local power relations, or the required sustainable character of development.

In addition to these changes in the content of rural development policies, the implementation of these policies through planning underwent gradual but important changes as well. In the early days of development planning, attention tended to be focused on macroeconomic planning at the national level. Gradually, however, planners realized that one of the reasons for the differences between what was planned and what actually happened was the failure to undertake more detailed planning of individual sectors, projects, and regions (Porter and Olsen 1976; Conyers and Warren 1988). The importance of sectoral planning and project planning was recognized earlier than that of regional planning and led to the strengthening of development planning capacities within sectoral ministries and to the improvement of project planning and management techniques. Over time, the role of planning approaches which take into account regional differences in development constraints and potentials, has been recognized (Conyers 1983; 1985). Various governments, such as Kenya, Tanzania, Indonesia, and Botswana, have adopted some form of regional development planning, which corresponds with the complex character of rural development.

Regional development planning may be defined as a horizontal planning approach for a subnational geographical unit (the delineation can either be based on administrative, physical, or functional motives), with the aim of supporting development. The horizontal perspective attempts to analyse and integrate the various aspects of development planning at one specific level. This may include such activities as economic planning, land use planning and social service planning, with the goal of improving intersectoral integration (Drusseldorp 1980);

Regional development planning and regional development plans may take many different forms. Variable levels of cooperation between and within sectoral ministries at the national and regional level may lead to significant differences in the quality of development planning efforts. Ideally, regional development planning starts with a comprehensive regional analysis of development resources and constraints which guides sectoral activities and the selection of individual projects. In practice however, regional development planning is often limited to a mere junction of sectoral plans. This last interpretation approaches Luning's concept of 'regionalized planning', which he defines as 'the adaption of sectoral plans to regional circumstances' (Luning 1981: 10).

The level of detail of regional development plans may vary according to local circumstances, such as available planning staff, resource position, and the sheer number of government projects to be implemented. Plans may range from an inception plan which presents a broad assessment of the major potentials and constraints for development and outlines the more important development activities, to a detailed development plan which contains programmes of action detailed to the level of identification of projects (Dusseldorp 1980).

Whatever form regional development planning takes, compared with sectoral planning, several potential advantages may be identified. These are: the capability to improve horizontal and vertical integration of sectoral activities; the facilitation of popular participation in planning; and the ability to increase the quality and relevance of planning due to its comprehensive character.

The integrative function of regional development planning appears in two ways. First, regional development planning provides for the opportunity to integrate sectoral planning into the spatial framework of the region. Through this, an improved coordination between the activities of sectoral ministries at the regional level may be achieved. Next to this, regional development planning may contribute to an improvement of the vertical integration of development activities. The regional level is the ideal level for integrating village projects into a regional framework following national goals. Also it is the level at which national plans are detailed into local plans based on a regional framework. In the ideal situation this would lead to an integration of the advantages of both a top-down and bottom-up planning approach.

Second, regional development planning may facilitate popular participation, as it is undertaken at a spatial level which is 'closer to the people' (Conyers and Hills 1984: 219). It provides for an incentive for involvement of area residents in the planning process. Participation is not only an end in itself within the structure of a 'democratic society', but may also be seen as a way of improving the quality and relevance of plans and as a means of facilitating their implementation and acceptance – especially with respect to rural development planning, because rural development plans are particularly susceptible to variations in local conditions, needs, and attitudes. Through popular participation, information can be obtained about these local variations which will enhance the quality and relevance of regional development plans.

Finally, regional development planning may increase the relevance and quality of development planning because ideally it starts from an overall regional analysis of development potentials and constraints, taking into account the physical, economic, and social aspects of the region; also, it may guide and support the activities of planning agencies and provides them with more relevant information. If supported by decision-making power at the same decentralized level it will lead to improved regional development.

However, integrated regional development planning is not a panacea to all planning problems, having its own drawbacks as well. Based on empirical studies in several African countries, Hyden (1983: 94) comes to the conclusion that: 'the creation of administratively integrated, multi-functional programmes in an administratively weak environment has proved a great mistake . . . if the operation of each programme component is made dependent on the others, there is a high probability that the whole programme will not work'. In my opinion this does not mean that a regional planning approach is inappropriate for Third World countries. What it points out, however, is the fact that regional development planning should be seen as complementary to a sectoral planning approach. Precautions have to be taken and any regional development planning structure has to be introduced step by step in a flexible way and adjusted to local circumstances concerning the quality of the administrative environment.

The country-specific character of regional development planning – measured in the degree of horizontal and vertical coordination, the extent of popular participation, and the level of detail of regional development plans – and its practical value for improved rural development depends to a large extent on the national planning environment, the regional differentiation of natural and human resources, and the organization of regional development planning and the character and degree of decentralized decision-making.

At the regional level, the following factors will determine the specific character of the regional development plans in each area: the regional planning environment; the quality and amount of planning staff; the quality and amount of data available for a regional analysis; and the actual participation of the population in planning. The amount of regional differentiation in the distribution of these factors will determine the differences in regional development plans within a country.

The relation between decentralized decision-making power and regional development planning in general needs clarification. Hyden (1983) defined decentalization as: a dispersal of decision-making power to individuals or institutions at lower levels in a hierarchy. Both character and degree of decentralization differ substantially between countries and within countries over the years. In order to describe these variations in character, the concept of decentralization is sub-divided into deconcentration or administrative decentralization, and devolution or political decentralization. In practice, a combination of both forms of decentralization is common. Deconcentration refers to the delegation of decision-making power to lower levels within the same organization, in order to improve the efficiency of the organization. For instance the delegated power of a district officer in the Ministry of Agriculture to allocate money to small agricultural projects. Devolution, in contrast, refers to an inter-organizational transfer of power to geographical units of local government outside the direct command structure of central government with the aim of increasing both the involvement of these levels in decision-making, as well as the efficiency of service delivery; for example, a district council which runs a primary education programme. In addition, the degree of transferred power and responsibility for planning and managing development activities to lower levels in a spatial hierarchy, which determines the extent of self-government, may vary substantially.

Decentralization and regional planning almost serve the same purpose. The potential advantages of both concepts are much alike, and are, if implemented properly, mutually supportive. Decentralization can be seen as the logical counterpart of regional planning in public administration. Depending on both character and degree of the transfer of decision-making power, decentralization facilitates the administrative implementation of regional development planning. The potential benefits of regional development planning may materialize sooner when supported by decentralized decision-making, especially in relation to the allocation of funds. It is this important relation between decentralization and regional planning that gives the concept of decentralization its spatial dimension.

Whereas regional development planning is the organizational form of planning through which popular participation may be incorporated in the planning process, decentralization provides the administrative structure which enables popular participation in planning and decision-making, provided substantial power is transferred to locally elected political representatives. However, decentralization is not simply a geographical relocation of the physical infrastructure of bureaucratic power. Slater (1989) and Smith (1985) have stressed that such a relocation can be perfectly compatible with the political centralization of that same power. According to Hyden (1983) and Samoff (1979), this is what actually happened in many African countries where decentralized administrative structure did not effectively reach beyond the construction of office blocks in sub-national government centres.

Yet, on the other hand, real devolution of decision-making power does not automatically lead to increased or effective popular participation. A great deal depends on the existing social and political structure in the area, and on the characteristics and methods of selection of the local political leaders (Conyers and Hills 1984). Local government might be highly decentralized but based on a small existing elite only. If so, it is possible that the interest and views of the mass of the population are better represented through central government field staff.

In addition to the potential advantage of decentralization regarding increased popular participation in regional development planning, decentralization of decision-making power may also create a supportive administrative structure for regional planning. Decision-making at the regional level, either by a locally elected body or by field officers, will be more likely than coordination between various agencies involved in planning and implementing development programmes. Especially if, as a part of the decentralization programme of a country, a development manager or a committee is introduced with formal coordinating power. Not only will the coordination improve, but also the flexibility and responsiveness of development planning to local conditions, which will support a regional differentiation of planning (Rondinelli 1981; Rondinelli and Cheema 1983).

The conclusion is that decentralization of decision-making power encourages popular participation and promotes coordination of regional development planning. However, the degree, nature, and success of both participation and coordination depends not only on the character of decentralization but also on national and local social and political structures, organization, and commitment.

THE CASE OF BOTSWANA AS THE OBJECT OF STUDY

Botswana is one of the most interesting countries in Africa in which to study the effect of regional development planning and the influence of decentralization of planning and decision-making power on rural development policy and regional development planning. It is so for several reasons. The extent of the rural development policy has been substantial compared with other African countries, regional development planning was introduced in the mid-1970s and is of a relatively high quality, and a certain amount of decision-making power has been decentralized to the district level.

Although government development investments have always been primarily directed towards the development of urban areas and prospective mining activities, rural development has received ever-increasing attention over the years. The sustained economic growth, which contributes to the financially strong position of the government, combined with substantial donor support, has been the financial base of this rural development policy.

Regional development planning (in the form of district development planning) was introduced in the mid-1970s. The quality of this development planning is high when compared to other African countries and Botswana has a relatively long planning tradition and a well-developed and qualified planning machinery, both at the central and district level. Because of its small population of approximately 1.3 million in 1990 (CSO 1987) and its simple administrative and political spatial hierarchy (Botswana knows only a central and district level of government representation), the management of development activities is relatively easy, due to small-scale activities and short lines of communication (Gasper 1987).

Since independence in 1966, the government has decentralized certain responsibilities and decision-making power to both field officers and democratically elected councils at district level. Although central government control over district councils has become stronger during the past twenty years, they still perform important political and administrative rural development functions. Though this period may seem rather short and the performance not exceptional, it is nevertheless remarkable when compared with other African commonwealth states which inherited the English-style local government. In countries like Ghana, Sierra Leone, Tanzania, Nigeria, Zambia, and Kenya, councils were, after an initial period of optimism, either suspended or stripped of their major functions (Tordoff 1988). Financial constraints, inadequate staffing, corruption and inefficiency, all restricted the ability of the councils in these countries to handle rural development problems. At the same time bureaucratic and political elites were reluctant to transfer meaningful power to these institutions, being afraid of their potential to become a platform for ethnic or regional political activity thus countervailing the power of these national elites. In practice, decentralization became limited to some form of deconcentration, leading to a closer fusion between central and local government and in fact reducing popular participation (Tordoff 1988; Hyden 1983).

RURAL DEVELOPMENT POLICY AND REGIONAL DEVELOPMENT PLANNING IN BOTSWANA

Rural development policy in Botswana

Examining rural development policy in Botswana from the eve of independence in 1966 to date shows a growing realization of the needs and rights of the rural population on the part of the government since the early 1970s. The emphasis has shifted progressively from social service provision to a more integrated approach with particular focus on the creation of productive income-generating employment.

Although the objective of rural development and the nature of strategies pursued have altered over the years, reflecting changing perspectives on the part of international donor organizations, two major aspects have played an important role in setting the direction of rural development – namely, the development of a growth economy based, apart from mining activities, on the cattle industry and the pursuit of political acquiescence in the rural areas (Picard 1987; Holm 1982).

The 1969 elections showed that the ruling Botswana Democratic Party (BDP) did not have full control over the rural areas. To pursue political acquiescence in the rural areas it became evident that the BDP would have to embark on an articulated rural development policy with visible and tangible results. A basic needs oriented programme, the Accelerated Rural Development Programme (ARDP), was initiated in November 1973, less than one year before the 1974 general election.

The ARDP was seen largely as a district-level activity, separated from central government operations (Picard 1987). It concentrated on providing social services and physical infrastructure to the rural population in the form of primary schools, health posts, road improvement, and water reticulation. Judged on its own criteria the ARDP was a success in that many projects had been completed or were well underway by the election date. The ARDP increased capital investment in the rural areas more than fourfold during the 1973–6 period of its operation, spending more than Pu20m (Colclough and Fallon 1983). Roughly 50 per cent of this came from domestic development funds (Egner 1978). The programme benefited greatly from the earlier strengthening of district development staff and from strong political commitment (Colclough and McCarty 1980).

Holm (1982: 85) investigated the politics of rural development until the early 1980s, concluding 'that rural development is a relatively unimportant consideration in the voters' decision'. This particularly reflects the aftermath of ARDP, or rather, what had been held in store by the country's political leaders prior to the 1974 elections, as the major direction for rural development had been laid down in a government White Paper in 1973. Entitled *National Policy for Rural Development*, this document clearly spelled out the future direction of rural development. It concentrated on a crucial land reform measure, namely the allocation of large tracts of tribal land to major cattle owners under exclusive tenure conditions (Botswana

1973). It was obvious that a major transformation of the rural areas would result from the pursuit of large-scale commercialization of the livestock industry through the privatization of land. However, the voters were only made aware of this in July 1975 when the Tribal Grazing Land Policy (TGLP) was announced.

Though infrastructural developments continued, the TGLP had a large impact on rural development; in the words of Picard (1987: 243) 'If the ARDP represented symbolism and political quiescence, the TGLP represented substance and economic transformation.' At the heart of the policy was the allocation of huge commercial 'Texas style' ranches with the aim of alleviating grazing pressure in the overcrowded eastern grazing areas. The TGLP became a 'dismal failure' (Sterkenburg 1987) as it led to the creation of several large ranches but not to better herd management; it did nothing to curb overgrazing in communal areas. Instead of reducing inequalities in the rural areas, the TGLP contributed to a further increase of them (Bekure and Dyson-Hudson 1982; Hinderink and Sterkenburg 1987; Sandford 1980).

The situation at the end of the 1970s was such that Holm lamented

> that Botswana's rural development programme receives minimal financing, shows relatively little concern with the mass of the population, is articulated by national rather than local officials, is heavily foreign funded and controlled, and yields few results which benefit the living conditions of the average voter.
>
> (Holm 1982: 86)

Within government circles, a realization came about some ten years ago, that small rural livestock owners and crop farmers were suffering substantially. Consequently, efforts were also directed to subsidization programmes in the field of water development, herd management, ploughing, and planting. The gradual change in rural development thinking from infrastructure and social service provision to productive employment creation is illustrated by the launching in the early 1980s of a Financial Assistance Programme (FAP) for the promotion of various rural industries. Another example of these changes in the character of rural development policies and planning strategies for rural development is the Communal First Development Area (CFDA) strategy, which was introduced in 1982 and may be seen as Botswana's version of an integrated rural development strategy.

These changes in rural development policy supported the need for a regional approach to development planning. Because of the highly intensive (requiring a lot of basic data material and individual contact between planners, extension workers, and the population) and mutually supportive character of the rural development policy nowadays, a regional planning approach became more and more necessary, a need which contributed to the early introduction of district development planning in the mid-1970s.

Regional development planning in Botswana

Regional development planning in Botswana is executed at the district level through both central and district level institutions which existed prior to its introduction in the mid-1970s. In order to comprehend the organization structure in which development planning takes place, a brief description of the government agencies at the national and district level involved in development planning is essential (see Reilly 1981; Tordoff 1988).

Central government institutions

Nation-wide development planning in Botswana is assigned to the Ministry of Finance and Development Planning (MFDP). This ministry is responsible for the National Development Plans (NDP) which outline national goals and priorities and cover a period of five to six years. MFDP houses the Rural Development Unit (RDU) which coordinates all rural development issues at the centre and performs the secretarial duties of the high-level Rural Development Council (RDC). The RDU has only a few executive responsibilities, but coordinates largely by working through informal relations and formal coordinating committees. The RDC is the key policy-making body concerned exclusively with rural development and chaired by the Minister of the MFDP. Besides this central function in the planning process, the MFDP is well represented within each ministry, as staff members of the planning units of all ministries are employees of the MFDP.

The other key ministry engaged in development planning is the Ministry of Local Government and Lands (MLGL). MLGL is administratively responsible for all district government institutions. It will, says the National Development Plan, 'work towards encouraging greater decentralization of decision-making' but also 'provide effective representation of Central Government at district level and co-ordinate local authority activities, through an effective District Administration [DA] machinery' (MFDP 1985: 77). MLGL has a major role in formulating rural development policy, organizing the district planning process through its District Plans Committee (DPC), and in coordinating, at the centre, many of the functional responsibilities of the local authorities. The MLGL houses, amongst others, the Department of Town and Regional Planning (DTRP) and the headquarters of the Unified Local Government Service (ULGS).

Most sectoral ministries (such as Agriculture, Education and Health) are also involved in district planning and are represented at the district level as well. Although the degree of deconcentrated power varies, most of them retain strong vertical links with the centre, where most decisions are taken.

District level institutions

There are four district level government institutions that fall under the responsibility of MLGL: the district administration, the land board, the tribal administration, and the district council (see Figure 6.1). The district administration consists of the district commissioner (DC), the district officers (DO), and supporting staff. The DC has little power compared with his colonial predecessor, because a great part of the authority has either been transferred to the district council or to field officers of sectoral ministries. Since independence there has been a gradual shift in the tasks of the DC away from political control and more towards the role of development manager. An example of this shift has been the appointment of the DC in 1971 as the chairman of the new District Development Committee (DDC), the development coordinating body at district level. Membership includes representatives of the district administration, of several important sectoral ministries at district level, the council secretary as co-chairman, the council planning officer and representatives of the district council, land board, and tribal administration. The function of the DDC is to serve as planning body and discussion forum for the district; to coordinate the work of the various central and local government agencies; to prepare and oversee the implementation of district development plans (DDPs) (Egner 1873), and to consult village development committees (VDCs) (the village level institution for the promotion of development), and extension staff based in the rural areas so as to ensure the incorporation of grass roots ideas into the DDPs.

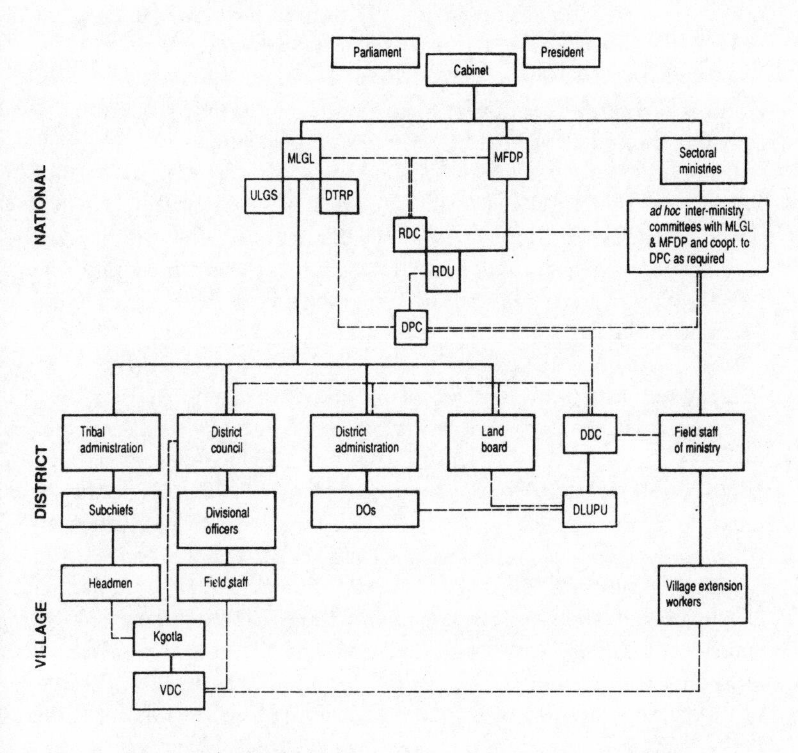

Figure 6.1 Botswana: the main government institutions concerned with district development planning
Source: Adapted from Reilly 1981 and MLGL 1979.

In the interest of efficiency and fair play, the right to allocate tribal land was transferred in 1970 from the chiefs to newly established land boards. These retained a semi-autonomous status in order to keep them independent from district council and political influences (Reilly 1983). The membership consists of representatives of the district council and tribal administration, as well as nominees by the MLGL. Next to the right to allocate tribal land, the land board is responsible for a variety of issues concerning district land use policy.

The tribal administration was severely curtailed at independence when the chiefs were stripped of most of their statutory powers. The main reason for the central government referring these powers to newly established district councils was to limit the power and potential threat of the chiefs and to create a local government body strongly controlled by the centre. Since then, the main official task of the chiefs is the administration of justice under the system of customary law, for which they are paid by the central government. In addition, the chiefs still are an important beacon in the lives of most rural people, and they fulfil a crucial role in people's participation in rural development because they are the only district institution that reaches most people on a regular base through the Kgotla meetings: the traditional gathering presided over by the chief or his representative (Tordoff 1988).

The district councils are the only elected bodies at district level. Whether and to what extent decentralization in the sense of devolution is a reality in Botswana depends on the autonomy and decision-making capacity of the district councils. They came into existence on 1 July 1966, three months prior to independence. Although established under colonial rule, they are seen as a post-colonial phenomenon, which may partly explain why they still exist in Botswana. It was initially thought that they would in time take over fully from district administration and most of the former responsibilities of the chiefs. As will be shown, this turned out otherwise, and the district councils still hold almost the same statutory responsibilities as in 1966. These include: the provision of primary education, basic health care, the maintenance of ungazetted roads, water supplies, and social and community development. The district council is also the endorsing authority for the DDP and annual plan produced by the DDC. For the execution of these tasks, the district councils are assisted by permanent staff, consisting of a secretary, treasurer, planning and departmental officers, and industrial workers.

The 1965 Ordinance is phrased in such a way that the district councils would be financially autonomous through receipt of their principal revenues for recurrent expenditures from local government taxes. Over the years they became more and more dependent on deficit grants from central government, which simultaneously increased financial control (Egner 1987). For their development expenditures, the district councils rely on domestic development funds for nearly half the total budget, and for the remainder on donor funds as well as borrowing from the World Bank (Tordoff 1988). Both financial control and the distribution of development capital over the districts and projects fall under the responsibility of MGL headquarters.

District development planning in Botswana

The decision to create a district development planning structure as it is known today in Botswana, was actuated by the ARDP which lasted from late 1973 to mid-1976. Through this programme, district councils became involved in rural development planning, and the need for structural coordination across sectors and institutional areas of responsibility at the district level and between districts and central government became necessary.

In 1977 the concept of the DDP was introduced in order to encourage all agents for development within each district to consider systematically their priorities and strategies, and to incorporate the wishes and needs of the population in development planning. The DDP would also analyse the implications for each district of the current National Development Plan, provide central government with a clear indication of the districts' choices of priorities and the recurrent manpower and finance implications of these choices, and to encourage coordination between planning agencies (Reilly 1981).

The district commissioner and the council secretary, as the joint plan managers, are responsible for all aspects of the DDP. In practice, most of the work is done by the district officer (development) (DOD) and the council planning officer (CPO). They should arrange the consultation of councillors, and, through cooperation with the chiefs, of the village development committees and people in the district through Kgotla meetings. They also arrange DDC meetings in order to bring together the plan contributors at the district level, and write most chapters of the plan. The format of the DDPs has only slightly changed over the years, and they still consist of the following parts: an assessment of the district, showing its needs, problems and potentials; an identification of priorities and broad strategies; the sectoral chapters under such headings as production and employment, social services, and physical infrastructure; the proposed development of the local institutions in the district; and the recurrent budget and manpower implications of the plan. It is the district council, as the ultimate representative body in the district, which gives final approval. As it is a district level plan, no approval from central government is required. A DDP cannot, on the other hand, commit central government to any future action or financial support.

Important for the implementation of the DDP are the annual plans (AP). These comprise a summary of the programmes and projects to be implemented during the year and, as such, constitute a plan of action designed to optimize the use of resources, encourage cooperation, and provide a means of monitoring progress. Also, the AP updates the DDP, as circumstances may change rapidly, thereby introducing a flexible and process approach element to regional planning.

Over the years the content of the DDPs has certainly improved, but they remain a reflection of the problems that regional planning encounters in Botswana. The problems which are closely related to the specific character of the political system and the decentralization process are described in more detail by Reilly (1981), Egner (1978, 1987), and Picard (1987), and may be summarized by the following three points.

First, participation of the district population in development planning hardly exceeds a pro forma consultation exercise. This consultation process is directed towards an identification of the population's needs for social services and physical infrastructure only, resulting in shopping lists from the villages. Most of them address insufficiently the structural problems of unemployment, land and draughtpower shortage, and the village's potentials for development. According to Noppen (1982) this specific character of the consultation process is partly caused by the paternalistic attitude of extension workers and planners, and partly by the undemocratic and elitist character of the Kgotla.

Second, the vertical integration between district and national development planning is still problematic. The essence of this problem is the fact that district plans do not receive full appreciation at central government level. Because there is no commitment of central government towards the implementation of the proposed programmes and projects, the link between planning and implementation remains weak. Budgets for local government institutions are normally available only one week before the start of the new financial year, and the actual financing takes place on a project-by-project basis, in order to keep strong control over the expenditures of local government (Egner 1987). Another important problem concerning this vertical integration is that until now the input of the DDPs in the NDP has been limited (Reilly 1981).

Third, sectoral development planning remains the primary form of planning, which results in DDPs that are often not more than an accumulation of sectoral plans. It is usually not the overall assessment of constraints and potentials in the district that guide and determine the content of the sectoral plans. More than anything else, they are the district's version of programmes and projects which have been introduced through ministerial headquarters in the capital (van Hoof and Jansen 1991).

BOTSWANA'S PLANNING ENVIRONMENT

The interaction between different agencies involved in development planning is shaped under the influence of and in reaction to external factors and processes which comprise the planning environment. Rondinelli and Cheema define the planning environment as:

> The specific and complex physical, socio-economic and political environment, that shapes not only the substance of policies, but the pattern of interorganizational relationships and the characteristics of implementing agencies, as well as determining the amount and types of resources available for carrying them out.
>
> (Rondinelli and Cheema 1983: 27)

An additional factor affecting the content of regional development planning and rural development policies is the disposition and distribution of physical resources. For the sake of clarity, this planning environment may be divided into four main elements: physical environment, population, economic structure and growth, and political system. Although the influence of these elements on the content of government action are not linear, they are discussed separately in the Botswana context.

The physical environment

Significantly, the physical environment sets the margins for government development, and especially rural development, efforts. The possibilities for rural development policy in Botswana are on the one hand restricted by its harsh climate and on the other hand enhanced by its relative abundance of mineral resources.

Botswana is bordered by Zimbabwe and Zambia to the east and north, and by Namibia to the north-west. A long southern boundary with South Africa completes this circle of countries surrounding the landlocked nation (see Figure 6.2). Most of the country is rather flat and constitutes a vast tableland at about 1,000 metres above sea level. Its total land mass is about the same as Kenya or France. The Kalahari desert is of major ecological influence in the country, covering about 75 per cent of the total area.

Botswana's climate is arid to semi-arid, with an average annual precipitation that ranges from 650 mm in the north-east to less than 250 mm in the south-west. Most rainfall takes the form of short, intensive showers during the summer, and fluctuates sharply in time and space. Drought is therefore a recurring hazard: since 1979/80 seven out of ten years have been registered as drought years. With the exception of the Okavango delta, surface water is scarce. Groundwater exists in most parts of the country and boreholes are an important source of supply.

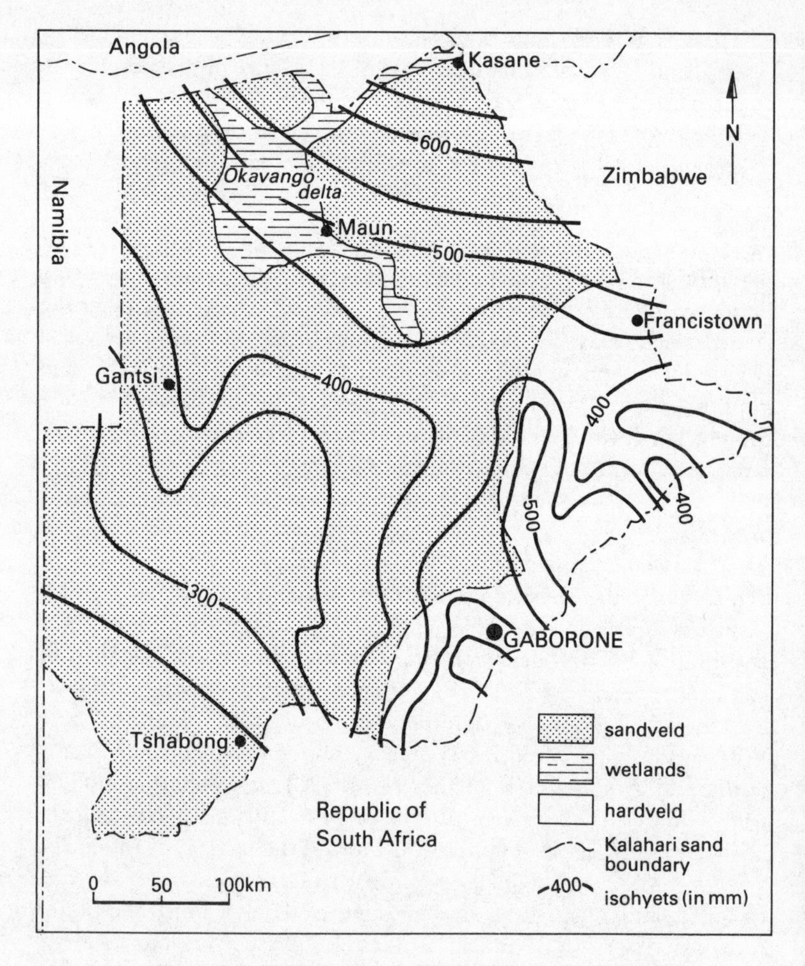

Figure 6.2 Botswana: rainfall distribution and soils
Source: Arntzen 1989: 56.

The regional differentiation in natural resources, of which the most important ones are precipitation and soil fertility, is great throughout the country (see Figure 6.2). Based on this variation, Arntzen (1989) distinguishes three ecological zones: the western/southern parts, where rainfall is lowest, soils are sandy, and surface water extremely scarce; the north, with the highest average rainfall (up to 650 mm per annum), large surface water resources, but with sandy soils of low fertility, and with mostly saline groundwater; and the east, with slightly more fertile soils, reasonable rainfall (400–500 mm per annum), and groundwater perspectives ranging from good to poor.

Arable agriculture is almost totally limited to the eastern part of the country, but even in this area it remains a rather precarious undertaking. A larger part of the country is suitable for extensive livestock keeping. Given Botswana's semi-arid climatic conditions and its fragile vegetation, the margins of intensified land use are limited. Unfortunately, the commercialization and growth of the livestock sector has led to cases of serious regional overstocking and environmental degradation (Arntzen 1989).

The restrictions of the physical environment have partly determined the character of the government's rural development policies. The support of the livestock sector during the first fifteen years after independence has been the easiest way to increase the contribution of the agricultural sector to economic growth. This perspective has been highly compatible with the interest of the cattle-owning political and bureaucratic elite. It also explains the recent interest in the development of non-agricultural activities in the rural areas and the increased paper attention given to resource management issues.

The regional differentiation in natural resources and environmental degradation have been, and will surely be in future, an important incentive for a regional approach towards development planning. Planners in Botswana progressively realize that a regional differentiation of development planning and an integrative approach towards rural development are needed in order to make optimal use of the limited resources.

Population

The total 1990 population of Botswana is estimated at 1.3 million which corresponds roughly with 2.5 persons per square kilometre. This is very low when compared with bordering countries like

Zimbabwe (approximately 24 persons per square kilometre) or Zambia (approximately 9.5 persons per square kilometre). Its annual growth rate of 3.4 per cent is, however, among the highest in the world. If current growth projections are accurate, the population will double by the turn of the century. Today, almost 50 per cent of the population is under 15 years of age (Granberg and Parkinson 1988). This implies a high dependency ratio, and puts great pressure on Botswana's government to provide for sufficient basic services. Large investments are needed to keep the quality of services like health facilities and education at the same level, not to mention the cost of a quality improvement of these services.

Due to the high population growth, labour-extensive agriculture, little industrial development, and a small formal sector, there is an excessive pressure on the labour market. Over 40 per cent of the 1984/5 labour force (estimated at 368,000) was unemployed or underemployed (CSO 1986). Every year the unemployed figure increases by another 9,500. The government of Botswana is increasingly aware of these problems and tries to create and stimulate new employment opportunities. This requires a planning approach adopted to the development constraints in Botswana. Because markets are limited and export is difficult, commercial activities will necessarily remain small scale. This implies that a planning approach is needed that gives good insight in the actual situation in the rural areas, and which is based on close contact with the people. For this sake, increasing attention is given to district development planning.

Another characteristic of the population of Botswana, necessitating a regional planning approach, is its highly distorted distribution (see Figure 6.3). Due to more favourable environmental conditions and a location along the historical trade route to the north, more than 80 per cent of the population lives in the eastern part of the country. The consequent range of rural population densities per square kilometre ranged from 17.2 in South East District to 0.2 in Ghantsi District in 1981 (Arntzen 1989) and require completely different district development plans.

Mitigation has been part of the Tswana lifestyle since the first settlement of these tribes in the region. Traditionally, migration had a seasonal character and took place between the homestead and the cattle post. During the twentieth century, seasonal labour migration to the Republic of South Africa (RSA) became the dominant form of migration. At its high point in the mid-1970s, 70,000 men (one third of the total labour force at that time) were working in the RSA

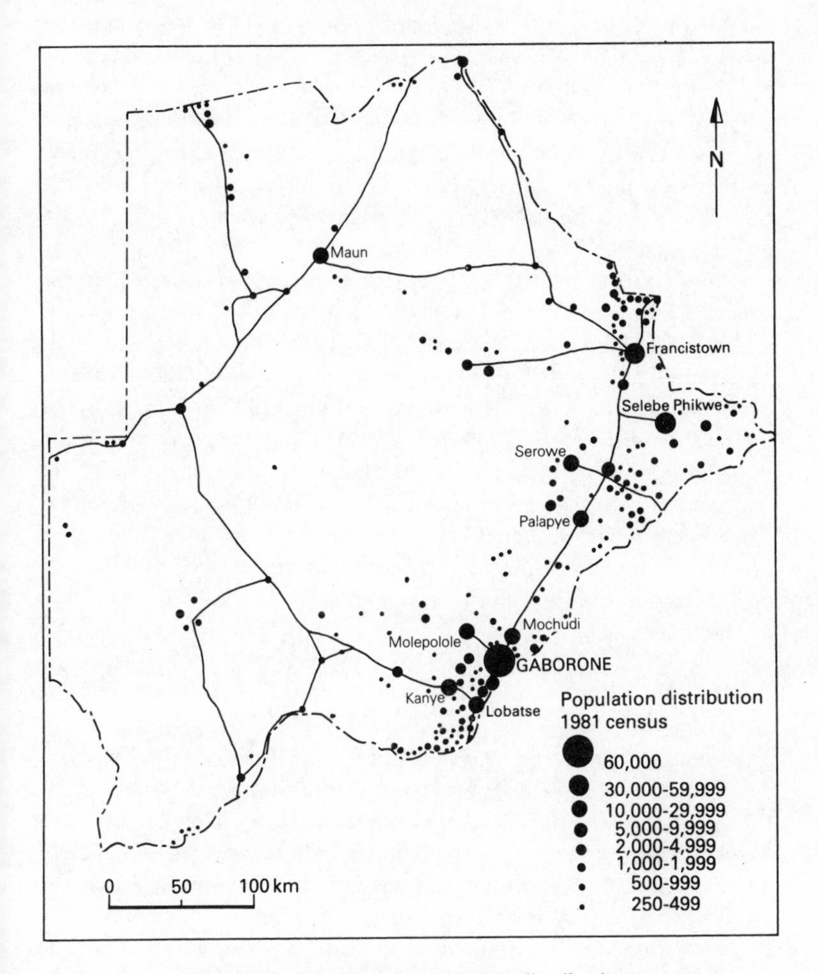

Figure 6.3 Botswana: population distribution
Source: MFDP 1985: 10.

(Colclough and McCarty 1980). Today, permanent migration of especially males to the urban areas strongly associated with the pattern of urban investment and job-creation since independence is the most prominent form of migration. Urbanization is therefore proceeding rapidly. Between 1965 and 1980 the urbanization rate was 12.4 per cent per annum, since 1981 it has been 8.1 per cent (World Bank 1989), which is high when compared with Zimbabwe (6.3 per cent) and Zambia (6.6 per cent), but this is also caused by the fact that it started from a modest base. In 1987 the percentage of the population living in urban areas had increased to an estimated 21 per cent (Zimbabwe 26 per cent and Zambia 53 per cent). This not only creates problems of unemployment, shanty town development, and increasing crime rates in the urban areas, but poses new problems for the rural areas as well. At this moment 65 per cent of the rural households are female headed. Related problems are unemployment and lack of labour force at the same time.

A final factor related to population and affecting government action, is ethnic composition. Compared with other African countries, though, Botswana has a relatively homogeneous population which has contributed to its stable political climate since independence, Various tribes do strive for a certain amount of self-control. The borders of the districts coincide more or less with tribal territories, and one of the reasons behind decentralization has been the wish of the tribes to determine at least part of the character of development themselves. Most Batswana do not see these wishes as a threat to national unity. A certain amount of self-control can contribute to genuine unity if based on the recognition and positive acceptance of regional cultural and ethnic diversities.

Economic structure and growth

Economic development, which is not identical to the much broader concept of development, is a necessary precondition if welfare gains for the majority of the population are to be made. Whether all levels of society profit from such growth depends on the ways in which these gains are distributed. Equally important are the ways in which the increase in income is generated, whether from employment intensive sectors such as arable agriculture or from sales of resources such as minerals.

At the eve of independence in 1966, the former British rulers questioned whether Botswana had any future at all. According to some of them, the best the country could aspire to was to become a 'functional bantustan' of South Africa (Morton and Ramsay 1987: 187). Although Botswana has remained highly dependent upon the Republic of South Africa, it did achieve an economic growth which is remarkable by sub-Saharan Africa standards (Sterkenburg 1987).

In 1966, Botswana was one of the poorest countries in the world, with a per capita income of Pu60 (then equivalent to approximately US$80) (Colclough and McCarty 1980). Five years of drought had struck the predominant rural population that scratched a precarious existence from subsistence agriculture. Almost twenty-five years later, affairs have partly been transformed. Seen from a macroeconomic point of view, the country is now doing well. Based on the profits of diamond mining, Botswana became financially strong, with an estimated capital reserve in foreign exchange equivalent to the cost of more than three years of imports (World Bank 1989). The government budget is in surplus and economic and social infrastructure has been created (Granberg and Parkinson 1988). Between 1966 and 1985/6, the GDP growth rate was around 10–12 per cent per annum (see Table 6.1), while GDP per capita rose from 240 to almost 1200 Pula.

Development however, is more than just economic growth. It also includes an improvement of the living conditions of the population and a greater security of existence for a large part of the population. The data in Table 6.1 say little about the living conditions of the rural poor – still the majority of the population of Botswana. For them, not much has changed over these years. Out of ten households, nine produce less than their subsistence needs. During periods of drought most of them rely on government relief support in order to survive. The rural income distribution survey, which was conducted in 1974, found that 45 per cent of the rural population had an income below the poverty datum line (Granberg and Parkinson 1988). The severe drought of the 1980s has certainly not improved this situation. Recent small-scale studies show an increase of the Gini coefficient from 0.52 in 1974 to 0.73 in 1987 (Granberg and Parkinson 1988), indicating an ongoing worsening of income distribution. This distribution of income is ameliorated, to some degree, by the provision of various public services which are free of charge, and of a quality and extent which are among the best in Africa.

Is Botswana the classic example of a dual economy, with a stagnating traditional agricultural sector and a booming modern sector with almost no links with the traditional one? Table 6.2 seems to support this statement, showing an important decrease in the proportional significance of the agricultural sector and an increasing importance of the diamond mining sector especially.

Although the picture of a dual economy does emerge, it oversimplifies the actual situation in Botswana. The traditional sector did not fully stagnate, whilst there are not insignificant links between the two most important sectors of the Botswana economy. As we will see, it is most of all the government, the fast expanding third sector, that provides for this link.

Table 6.1 Botswana: growth of Gross Domestic Product, 1966–1985/6

	1966		*1971/2*		*1975/6*		*1979/80*		*1983/4*		*1984/5*		*1985/6*
Total GDP (Pu million)	129		287		453		702		1,106		1,171		1,316
Average % p/a growth		15.6		12.1		11.6		12.0		5.9		12.4	
GDP per capita (Pula)	240		480		620		800		1,080		1,100		1,190
Average % p/a growth		13.4		6.6		6.6		7.6		1.9		8.2	

Source: Granberg and Parkinson (1988).

Table 6.2 Botswana: changes in the composition of Gross Domestic Product, 1966–1985/6

Sector	*1966*	*1971/2*	*1975/6*	*1979/80*	*1983/4*	*1984/5*	*1985/6*
Agriculture	39.4	32.3	24.0	11.1	6.0	5.2	3.9
Mining	—	10.9	12.3	30.0	31.7	36.3	46.9
Manufacturing	7.9	5.0	7.6	4.2	6.4	6.1	6.0
Water and electricity	0.8	1.3	4.1	2.1	2.5	2.1	1.9
Construction	5.7	9.7	6.9	5.2	5.5	3.9	2.9
Trade	18.5	17.0	15.5	22.4	21.7	20.5	17.6
Transport	8.1	3.7	4.6	1.9	2.6	2.5	2.0
Services	6.5	8.6	6.8	9.1	7.9	7.1	5.6
Government	13.3	11.5	14.7	13.2	15.8	16.2	13.3
Total	100.0	100.0	100.0	100.0	100.0	100.0	100.0

Source: Granberg and Parkinson (1988).

During the first ten years after independence especially there was a strong recovery of the agricultural sector, based on favourable weather conditions. The total output increased from Pu11m in 1965 to Pu86.6m in 1978/9, but drops back to Pu48m in 1985/6 due to drought (Colclough and McCarty 1980). It has been the expansion of the national cattle herd and the commercialization of the livestock sector, supported by lucrative export quota of chilled beef on EC markets at prices above those obtainable internationally, that has caused this growth. At this moment, it accounts for 80 per cent of the gross output in the agricultural sector (Granberg and Parkinson 1988). Because the ownership of cattle is highly skewed (7 per cent of the cattle owners own more than half of the cattle herd), the benefits of this growth have gone to a rather small number of large cattle owners. For some, these revenues have been the economic base of the continuation, and expansion, of their traditional political power.

Although the growth of the livestock sector has been significant, it is fully overshadowed by the expansion of the mining sector. In 1985/6 this sector accounts for almost half the GDP and for more than 55 per cent of the government revenues (see Table 6.3). The fact that it has been this capital intensive economic sector and not the manufacturing or arable agriculture sector that has been the vehicle behind Botswana's economic growth, is, for several reasons, of great importance for the place of the government in society and for the character of its policies.

Table 6.3 Botswana: central government budget, 1973–1986/7

	1973/4	*1975/6*	*1979/80*	*1983/4*	*1984/5*	*1985/6*	*1986/7*
A. Pula million at current prices							
Revenues and grants	48	88	249	563	803	1,133	1,424
Expenditures and net lending	61	90	228	460	615	719	1,011
Budget surplus	– 13	– 2	21	103	188	414	413
B. Percentage distribution							
(percentage of total revenue and grants)							
Mineral revenues	11	27	31	34	47	51	55
Customs revenues	44	28	32	28	19	13	14
Grants	13	9	15	9	5	4	2
Other revenues	32	36	22	29	29	32	29
Total revenues and grants	100	100	100	100	100	100	100

Source: Granberg and Parkinson (1988).

First of all, most of the revenues from this sector either go to foreign-owned companies or directly to the government of Botswana, which saw a tremendous increase in its budget during the last twenty years. Because of this, no substantial new economic elite arose to challenge the political power of the traditional elite. These revenues enabled the government to build up an extensive administrative and planning machinery. Compared to other developing countries, it also enabled the government to formulate a rural development policy which was relatively autonomous from the pressure of other governments or non-governmental organizations.

Second, much more than in other African countries, the state is not only an important political institution, but also an important economic force. In 1986 for example, 39 per cent of the formal sector wage employees were public servants (Granberg and Parkinson 1988). Given the fact that the internal private sector is only marginally developed and depends heavily on investments from the RSA, much of the internal economic investment is done by government.

After independence, the state power and pattern of intervention expanded as the independent government maintained the colonial monopoly of control over the economy in the absence of a strong indigenous capitalist class. Picard (1987: 10) speaks in this context of a 'state managed private sector economy'. Because a large part of the National Income flows through the hands of civil servants, government plays an important role in the distribution of welfare and in the creation of employment. Goverment planning is therefore of great importance, which partly accounts for Botswana's overdeveloped bureaucratic machinery.

A final consequence of Botswana's economic growth being mainly based on only one sector is the vulnerability of both the economy and the position of the government – the former being dependent on world market prices and the latter on events outside its own control; Egner (1987: 101–2) gives an example of the effects of this vulnerability. A temporary reduction in diamond prices at the world market from September 1981 to March 1982 cut back the foreign exchange reserve to half its former value in only a few months. The Ministry of Finance and Development Planning immediately imposed a 20 per cent across-the-board cut in all forms of government spending, and sharply reduced the development budget for district councils for the following three years.

The era of political stability in Botswana, made possible by this economic growth, may end as soon as revenues diminish. If no new diamond sites become economically exploitable in the near future, or no other economic sectors take over the leading role of the mining sector, economic growth will likely stagnate at the turn of this century.

Until now, the government has tried to attract new businesses through measures like financial assistance programmes or favourable tax laws in order to bring about a diversification of the economy. Several factors, however, impede such a development (MFDP: 1983). Botswana has a limited resource base and a small internal market dominated by businesses and products from the RSA, who can freely enter Botswana's market. Because its landlocked position creates high transport costs, and a dependence on RSA transport lines, export-oriented businesses cannot compete successfully. A penetration of the markets of neighbouring countries encounters difficulties, in spite of the intentions laid down in a regional economic cooperation scheme: the South African Development Cooperation Conference (SADCC).

The political system

A final element of the planning environment that is of great influence on the character and content of regional development planning and rural development policy is Botswana's political system.

The contemporary political system in Botswana is an amalgamation of pre-colonial traditional patterns of authority, colonial administrative structures, and post-colonial corporatist influences (Picard 1987). British colonial rule has been important for the contemporary character of government in Botswana. It introduced the administrative state form, which gives great allocative power to administrative elites, and provided for a blueprint for the present local government structure (for further detail see Picard 1987).

It has been the flexibility of a group of families of the traditional elite that secured their power already before official independence. Anticipating on the continuing importance of the bureaucracy, the so-called 'new men' (the well educated members of the ruling families of the traditional chieftainships) took over (informally supported by the British rulers) the key positions in the civil service, and adopted the system of administrative government and strengthened it after independence.

It was this group of men with both traditional and professional status who formed the BDP in 1962 under the leadership of Seretse

Khama, the chief of the Bamangwato, the biggest tribe in Botswana. With the non-official support of the colonial administration, the organizational back-up of the bureaucracy (which turned out to be a long-lasting alliance), and the financial assistance of the South African ranching community, the BDP had already secured its leading position before the first elections in 1965.

Although Botswana is a multi-party democracy with free and open elections, its functions *de facto* as a one-party state. Without the ability to use the bureaucracy as an overt mechanism of political mediation and mobilization (Picard 1987), and without the support of local notables and sufficient financial resources, the opposition parties (the Botswana National Front (BNF) and the Botswana People's Party (BPP)) have been unable to become a real threat to BDP power. The direct links between land and cattle owning and political and bureaucratic elites have been extended over the years, making these elites the main beneficiaries of economic growth and government policy.

These close links between the national elites and the absence of a strong opposition, may be seen as the two most important characteristics of Botswana's political system. It explains why, during the 1970s, a rural development policy could be formulated and implemented that turned out to be contradictory to the interests of almost 90 per cent of the rural population. It also explains the contemporary overemphasis in rural development policy on social service delivery and infrastructural works, because by implementing such a policy the BDP secures the rural voters' support without being obliged to formulate a policy that would structurally change the elitist character of the rural society.

Because of the intimate links between the elites, a perfection of the administrative style government has occurred. Several authors (Egner 1987; Gasper 1987; Picard 1979, 1987) agree with Gunderson (1971 in Picard 1987: 13) when he calls Botswana 'the closest current approximation to Weber's "administrative state"', by which he means that politicians determine the boundaries of government action and leave a high proportion of policy definition and routine allocation to their bureaucrats (Picard 1987). According to Holm, this results in a situation where policy debate takes place within ministries, and not in parliament or in public discussion. Dominant ministries shape the content of policies, while elected officials (both at central and district level) operate as reactors and can only (in the last extremity) veto proposals (Holm 1985). These central government bureaucrats exercise more authority over elected bodies (such as councils) than strict law or sound practice should allow.

Another characteristic of this administrative style of government that hampers further improvement of the district development planning approach is its preference for sectoral and hierarchical (top–down) structures. In practice these structures turn out to be major constraints for both a horizontal and vertical integration of development planning activities. It also explains why a meaningful devolution or even deconcentration of decision-making power to the district level – which is a necessary precondition for both forms of integration – has not occurred until now. To explain this, a closer look at the decentralization process in Botswana is necessary.

DECENTRALIZATION IN BOTSWANA

The decentralization process in Botswana since 1966 may be characterized as a decentralization of resources, which is balanced by a centralization of control. Or, as Egner (1987: 47) says 'administrative solutions to local authorities' problems have been preferred, rather than political solutions which give local authorities clearer authority'.

Immediately after independence, it became clear that demands for projects far exceeded the district councils' financial and managerial capacities. In particular, the MFDP became increasingly sceptical about the ability of the district councils to act. This negative attitude has lasted over the years, and turned out to be the main force behind the centralizing tendencies of central government. Certainly at the beginning of the 1970s a total dismantling of the district council system was more than once considered. In 1970 the Office of the President decided to strengthen the rule of the district administration at the expense of the district councils, through the establishment of the DDCs under the supervision of the district administration. Because of the weakness of the district councils, it was felt that the central government officials, both in the field and in central ministries, should closely supervise the planning and administration of all rural development projects (Picard and Morgan 1985). For this purpose the post of district officer (development) (DOD) was added to the DA. This caused many councillors to complain about the DDC taking over the council's responsibilities and reducing it to the status of an advisory body.

In addition to this, the ULGS was established in 1973 at the MLGL, which meant a central take-over of the hiring, transfer, and training of middle and senior level council staff. Not only are councillors now politically responsible for the activities of their employees over whom they have no proper authority (Egner 1987), they are also isolated from the lines of communication between the centre and periphery because these lines are mainly developed within the administrative structure of the state (Picard 1987). A council secretary charged that: 'we have become a part of bureaucracy. This gives all the power to the civil servants; none to the politicians' (quoted in Picard 1987: 195).

At this stage, MLGL that if no measures were taken, 'what had happened in Kenya (loss of council authority to a strengthened DA), would inevitably happen here' (Egner, quoted in Picard 1987: 181). From 1972/3 MLGL started to address the staffing and budgetary problems of the councils. Through this, councils were administratively strengthened, but at the same time this meant a further reduction of their political autonomy, as both DDC and central government had to approve expenditures.

In 1978/9 a Presidential Local Government Structure Commission was created to review the structure of all institutions of local government. Its main recommendation was to stop the trend towards more central control and to return a considerable amount of autonomy to locally elected bodies. Central government, however, rejected all major recommendations in 1981 and continued to extend its control over the district councils.

Although there has been a steady increase in the development and recurrent budgets of districts councils, and an improvement of the quality and amount of civil servants at the district level, central government's paternalistic attitude and financial control over the districts' expenditures have become stronger. Development funds are allocated on an annual basis and need a project-by-project approval by either the MLGL or the line ministry headquarters concerned (Gasper 1987). This affects the implementation rate and the relevance of the DDPs and annual plans negatively, reinforcing the prejudice at the central level that the districts are incapable of handling their funds in an efficient way.

Egner (1987) found that it was not a lack of capacity of local government that caused the disappointing implementation record for local government projects during the Fifth National Development Plan (1979–85), but that it was mainly budget blockage and lack of support in the MLGL and the MFDP headquarters that caused these delays. A final example of the increased monetary control is the abolishment of the Local Government Tax in 1987, thereby reducing the councils' own spending budget and making them almost entirely dependent on central government deficit funding (van Hoof and Jansen 1991).

Picard (1979, 1987) has tried to explain these trends in decentralization. According to him, the dynamic balance of power between central and local government has to be maintained as a precondition for councils to remain active. Dynamic, because changes in the planning environment – such as a reduction of financial resources – lead to adjustments of this balance of power. The sudden tightening of control by the MFDP and the non-acceptance of the DDPs in 1983 (caused by the temporary reduction in diamond income already mentioned) serves as a good example. In Picard's opinion councils are allowed to exist, and may have opposing political views because their function is largely symbolic and rather irrelevant to the decision-making process (Picard 1987).

This means that a further political decentralization is impossible. Councils will have to remain structurally weak because an increase in power would disturb the existing balance too much. As Picard and Morgan also note:

> For the dominant socio-economic elites, a potentially articulate system of local government . . . could come under the control of an opposition group which could challenge the economic interests of the ruling Democratic Party and its allies in the senior levels of civil service. The context of political and administrative structures, thus, requires a more centralized authority and this context is incompatible with a framework of autonomous local institutions.
>
> (Picard and Morgan 1985: 156)

It seems that the present relationship between central and local government is the only possible balance between keeping a democratic and pluralistic style of government alive without transferring real power to democratic institutions that could become a threat to the *status quo*. The margins for changes are very small, which explains why a decentralization of resources has been balanced by a centralization of control. Slater's conclusion concerning the decentralization process in Tanzania seems applicable to Botswana as well:

> A geographical re-location of the physical infrastructure of bureaucratic power can be perfectly compatible with the political

centralization of that same power. . . . Decentralization extended and consolidated the regional and local power of the state.

(Slater 1989: 515–16)

CONCLUSIONS

In this chapter I have described the influence of several processes at the national level on the character and content of rural development policy and district development planning in Botswana. The main reason for central government adding a regional development planning approach to the sectoral one has been the wish to increase the efficiency of government service delivery. The reasons for the introduction of district development planning were mostly technical: namely, the uneven distribution of both natural resources and human population over the country, the increased complexity of the rural development policy, and the involvement of more and more government institutions in development planning.

In order to achieve this aim, the government also improved the implementation and planning capacity of the district councils over the years; compared with other African countries, district councils in Botswana do fulfil an important role in government service delivery.

In the absence of politically articulated forces that worked for district self-government, central government has retained strong central control over the district councils. A further improvement of district development planning is only possible, in my view, if more decision-making power is also transferred to the district level, preferably through a devolution of power to the district councils.

What I have tried to show in this chapter is that the problems of district development planning at this moment are a logical conseqence of the choice not to devolute or even deconcentrate meaningful decision-making power. Without such deconcentration an increase of either vertical or horizontal integration of development planning activities, and also of the participation of the population in development planning, will not occur. As Picard showed, however, such a transfer of power is not very likely, given the vulnerable character of the balance of power between central and local government that exists at this moment in Botswana.

7

TECHNOLOGY ADOPTION AND POST-INDEPENDENCE TRANSFORMATION OF THE SMALL-SCALE FARMING SECTOR IN ZIMBABWE

Lovemore Zinyama

INTRODUCTION

According to Farmer (1981) the term 'Green Revolution' was popularized in the late 1960s to describe the consequences of the introduction in South Asia of improved agronomic packages involving new high-yielding varieties (HYVs) of cereals in association with chemical fertilizers, insecticides and pesticides, controlled water supply (usually involving irrigation), and new methods of cultivation. The adoption of this package of practices raised the possibilities for greatly increasing agricultural output, particularly for wheat and rice, to such an extent that famine and hunger would be banished forever.

In more recent years, we no longer refer to these changes in the rural landscape and economy as the Green Revolution, particularly in the context of Africa. Other, agriculturally less emotive but perhaps politically strong, terms such as agrarian change, rural or agricultural transformation, or simply agricultural and rural development are used. Whatever term we choose to use, the aims of many African governments today are broadly similar to those of the plant-breeding scientists in Mexico and the Philippines who developed the HYVs two decades ago, and to the governments of Asia who allowed their propagation and dissemination. African governments have finally come to realize the futility of their earlier post-independence development strategies that emphasized industrialization at the expense of agriculture and the rural sector. There is a general consensus now that the key to Africa's current economic crisis lies in the transformation and development of agriculture, the strengthening of the smallholder farming sector, the return to domestic food self-sufficiency, and the general development of the rural areas in order to make them more attractive places in which to live and work. However, one also needs to differentiate between intent and actual performance. Many African governments have so far failed to implement policies that can effectively tackle these problems, a major reason being their failure to find a balance between the frequently conflicting interests of their more articulate and politically better organized urban constituencies on the one hand and those of the politically weak rural population on the other. For instance, one prerequisite for getting agriculture moving again is to raise prices paid to farmers. Yet food price hikes have frequently caused urban riots in several countries in recent years, forcing their governments to withdraw the increases.

The aim of this chapter is to examine the adoption patterns for yield-increasing technologies by black small-scale farmers[1] in Zimbabwe during the first decade of independence in pursuance of the government's objectives to develop the previously neglected rural areas and to achieve national food self-sufficiency. Research in Zimbabwe reveals that black farmers rapidly adopted hybrid seed varieties and chemical fertilizers in the early 1980s as part of the process of transformation of the small-scale farming sector. The adoption of the yield-increasing technologies could be viewed as one indicator of an incipient agricultural or Green Revolution currently taking place in rural Zimbabwe. However, although a majority of farmers now use these new technologies, disparities in output, crop sales, and on-farm incomes between households remain. These variations are not due to differences in technology adoption alone, but are related to other socio-economic factors such as the distribution of farming assets, household demographic structure, and the gender and age of the household head.

The remainder of this chapter is divided into three sections. The following section briefly outlines the magnitude of the expansion in output and marketed qualities since 1980 of the principal food and cash crops grown by black small-scale farmers. The next section discusses the types of innovations and the trends in their adoption by small-scale farmers in the communal areas. The final section examines other socio-economic factors that have influenced the expansion of the small-scale farming sector during the past decade.[2]

CROP PRODUCTION TRENDS IN THE SMALL-SCALE FARMING SECTOR[3]

The performance of the post-independence government in Zimbabwe during the past decade in developing the small-scale communal agricultural sector makes it one of the few success stories in Africa today. When the new government came to power in 1980, it sought to reduce the racial and spatial inequalities in development that were the legacy of ninety years of white minority rule. From the outset, the government has given high priority to the development of the previously neglected subsistence communal farming areas that are home for some 60 per cent of the country's total population. Here, the black small-scale farmers have for the past century eked a living under conditions of a deteriorating man–land ratio arising from rapid population increase and widespread environmental degradation. Arable land holdings now comprise only about 2–4 hectares per farming household in these areas.

Since 1980, considerable progress has been made by the government in improving agricultural output in the communal areas, in increasing their share of marketed crops, and in raising on-farm incomes as well as generally improving living conditions through the provision of health facilities, water and sanitation, transport and communications infrastructure, and educational and other social services. For their part, black small-scale farmers have greatly increased their share of national agricultural output, particularly for maize and cotton their principal food and cash crops. In the decade before independence, maize production in the small-scale farming sector was largely static (see Figure 7.1). The sector contributed less than 10 per cent of the maize and less than 25 per cent of the seed cotton sold annually to the two parastatal marketing organizations: the Grain Marketing Board (GMB) and the Cotton Marketing Board (CMB) respectively. Today, it accounts for between 40 and 50 per cent of the maize and as much as 60 per cent of the cotton sold annually. Deliveries of maize from black farmers increased from a pre-independence peak of 84,265 tonnes sold after the 1976/7 harvest to a record of 819,000 tonnes following the 1984/5 season (see Figure 7.2). Seed cotton sales from the sector have increased fourfold from less than 45,000 tonnes annually before independence to a peak of 194,656 tonnes delivered after the 1987/8 harvest (see Figure 7.3). Such expansion in both output and sales is remarkable, especially when it is viewed against a backdrop of recurrent droughts during the

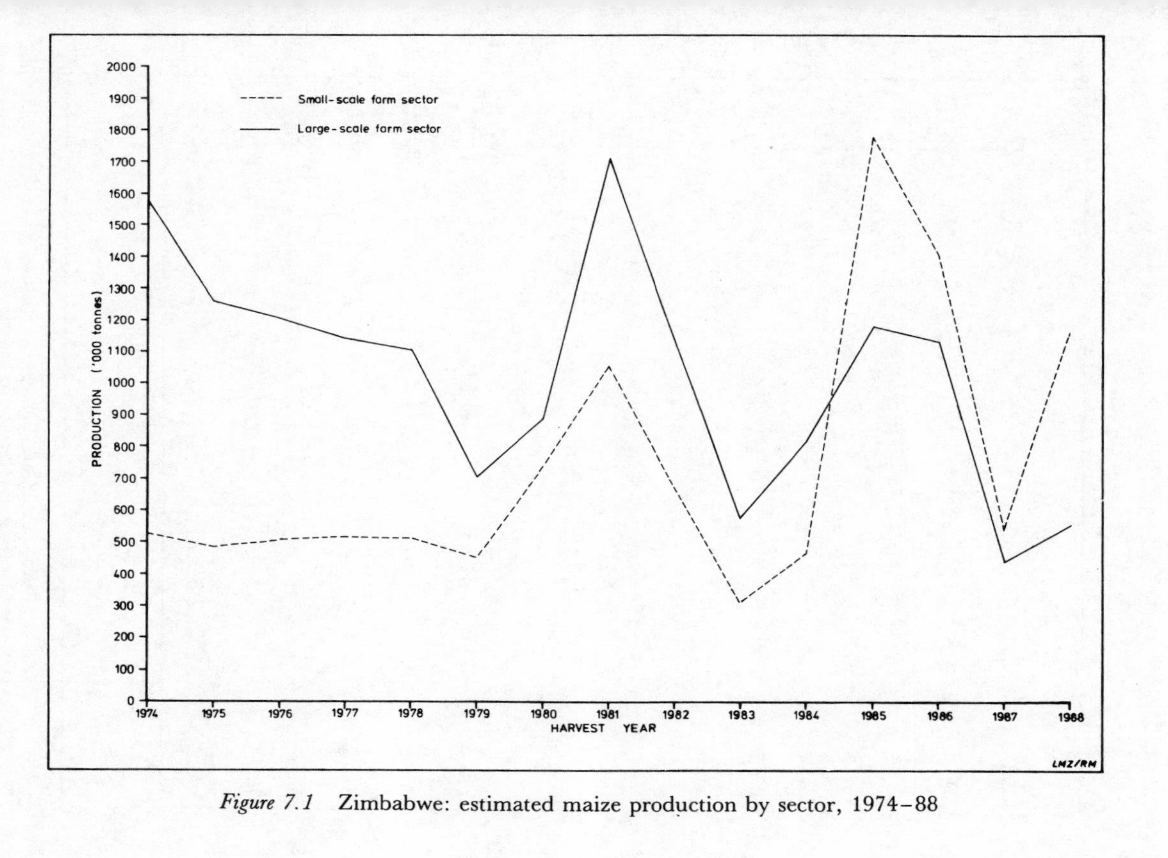

Figure 7.1 Zimbabwe: estimated maize production by sector, 1974–88

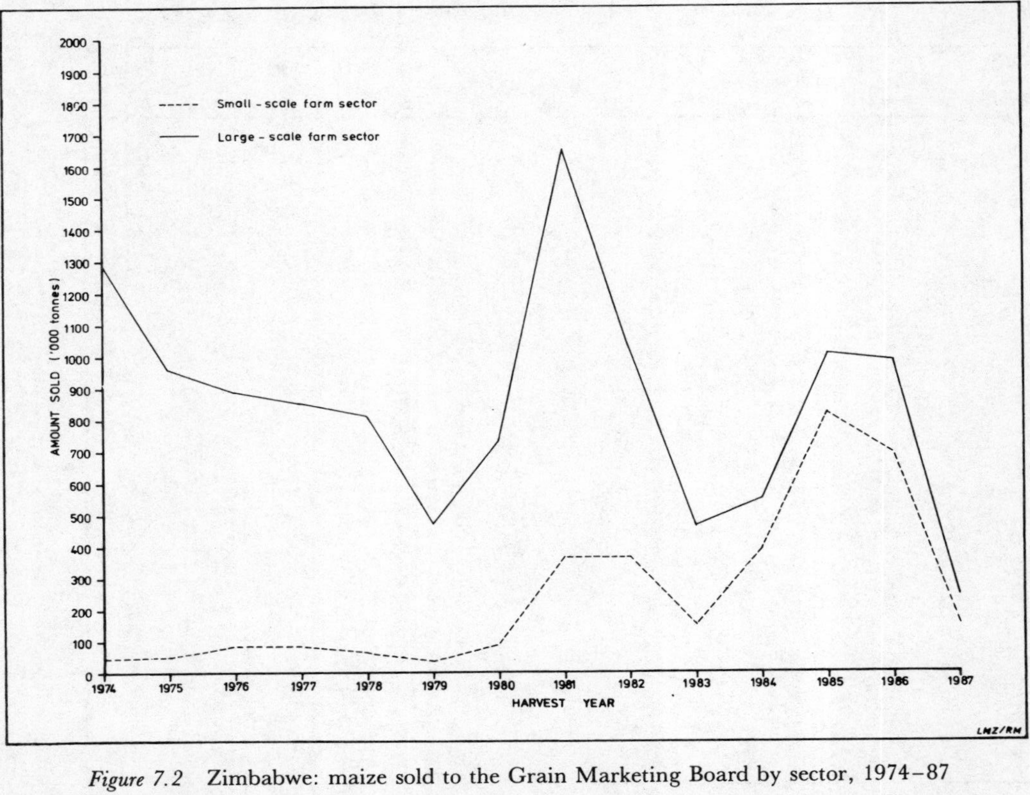

Figure 7.2 Zimbabwe: maize sold to the Grain Marketing Board by sector, 1974–87

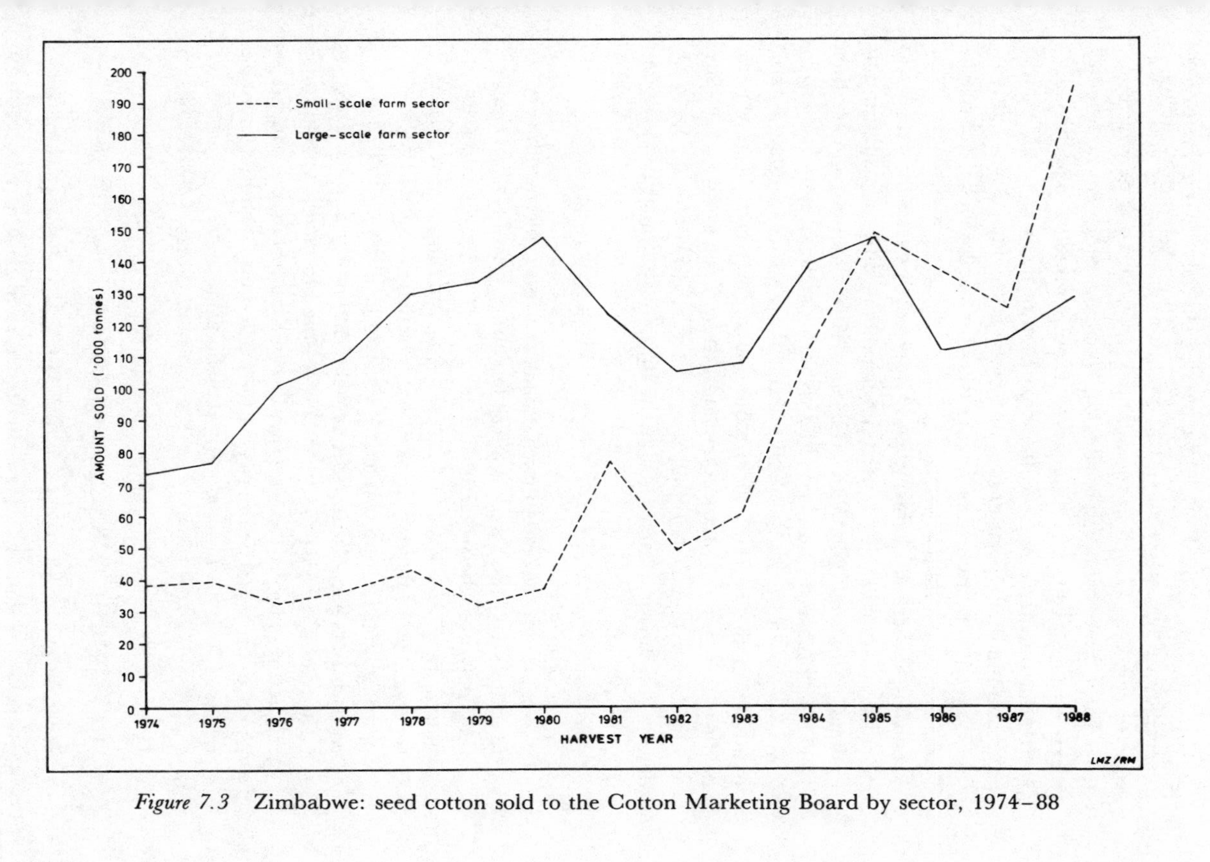

Figure 7.3 Zimbabwe: seed cotton sold to the Cotton Marketing Board by sector, 1974–88

mid-1980s and the generally poor performance of the agricultural sector and worsening food shortages over much of the African continent. A combination of post-independence state intervention and strong institutional encouragement, generally favourable prices (especially in the early 1980s), the provision of marketing and other support services, as well as the adoption by farmers of new farming techniques, have been responsible for this expansion. An additional short-term factor for the initial gains in output and sales in the early years after independence was the expansion in the area under maize cultivation in many parts of the country by both old and new cultivation, as well as by returning former war-displaced people (Rohrbach 1987; Zinyama 1986).

However, there is a growing body of research data which suggests that these nationally aggregated development gains hide substantial regional and social inequalities (Callear 1984; Chigume and Shaffer 1989; Gobbins and Prankard 1983; Rohrbach 1987; Stanning 1987; Zinyama 1988a,b). At the village level, these social inequalities are manifest on the basis of socio-economic status (there is as yet no distinct class differentiation), age and gender. The growth in output within the small-scale farming sector in Zimbabwe during the past decade and the associated social inequalities have similarities with the experiences of countries in Asia that went through the Green Revolution two decades ago. As noted by Farmer:

> bigger farmers *do* benefit most, smaller farmers (often with very tiny holdings) less, and landless labourers least of all (indeed they may even be worse off). The reasons are not far to seek, though they vary with crop, local social and land tenure structure, and other factors; for the bigger farmers are wealthier and more influential, and can afford, or acquire through influence, costly inputs like fertilizers, tube wells, and power-pumps. They have readier access, too, to credit, to the bureaucracy, and to marketing. Hence, far from the 'Green Revolution' ushering in an era of universal prosperity, as the earlier propagandists maintained, it can lead to widening income disparities.
>
> (Farmer 1981: 206)

AN AGRICULTURAL REVOLUTION IN ZIMBABWE?

Grigg (1984) has suggested four possible ways of defining the term 'agricultural revolution'. First, the term could be used to describe a radical change in the tenurial and structural character of agriculture. Second, it could describe the accelerated adoption of new farming methods. Third, it could be the accelerated rate of growth of total agricultural output or food output. Fourth, it could refer to an increase in the rate of productivity in agriculture, measured either by total factor productivity growth, the growth of total output per hectare, ot total food output per head of population. The overall impact of these changes on national development would be the provision of relatively cheap food based on low-cost, high-productivity agriculture and, ultimately, the transformation of the rural sector as occurred in Asia during the 1960s and 1970s and in Europe even earlier (Blackie 1987; Grigg 1984). These criteria will be discussed within the framework of the changes that have taken place in the small-scale communal farming sector of Zimbabwe during the past decade.

Tenurial and structural change

In Zimbabwe, black farmers living within what are now known as the communal farming areas (formerly African Reserves or Tribal Trust Lands) hold their land under a communal system of land tenure. Under this system, land belongs to the clan or group and individual families enjoy usufructuary rights only (Yudelman 1964). A family that is entitled to land in a given area will cultivate a piece of land on an individual basis as well as having access to communal grazing for its livestock. This basic structure of the communal system of tenure has remained relatively unchanged over the past century in spite of considerable pressures from, on the one hand, mounting population densities (Whitlow 1980; Kay 1976; Zinyama 1988a; Zinyama and Whitlow 1986) and, on the other, government attempts during the 1950s to introduce a more individualistic land tenure system that would have made land a market-exchangeable commodity (Floyd 1959; Yudelman 1964). Another colonial innovation was the establishment from the 1930s of seperate small-scale commercial farming areas (formerly known as African Purchase Areas) where few blacks with the means could purchase land under freehold or leasehold tenure. However, the number of people in these areas has always been very small. At the 1969 census, these small-scale commercial farming areas had 136,000 people or only 2.8 per cent of the total population. The subsector comprises some 8,500 farming units with an average size of less than 150 hectares: a total of 1.5 million hectares or 3.6 per cent of the national land area.

The post-independence government has categorically rejected proposals to change the communal land tenure system and to substitute it with individual freehold tenure. The reasons include the danger of worsening the problems of landlessness and rural indebtedness, and to avoid the emergence of a kulak-like rural capitalist class and increased social inequalities (FAO 196; Ministry of Local Government 1987). Instead, the current emphasis is on the reorganization and rationalization of rural land uses and village regrouping. A number of pilot projects have already been implemented in several parts of the country. By August 1987, the programme was in varying stages of implementation in fifty-nine villages covering 4,321 residential stands in different provinces throughout the country (Republic of Zimbabwe 1987). The rationalization programme is intended to provide for more efficient utilization of increasingly scarce land resources, to curtail the problem of environmental degradation that is being aggravated by the unplanned extension of settlement and cultivation into grazing and other marginal areas, and to facilitate the provision of social and economic infrastructure (Republic of Zimbabwe 1986; Ministry of Local Government 1987). It involves the consolidation of dispersed settlements into nucleated villages, the demarcation of arable lands, and their separation from communal grazing areas. The villagers are also encouraged to fence their grazing areas into paddocks in order to improve the quality of their pastures and enable them to practise rotational grazing and, generally, to improve their livestock husbandry methods.

The other structural change that has a direct impact on the population of the communal areas is the government's programme to redistribute land from the large-scale (formerly white) commercial farming sector and its resettlement with small-scale black farmers from the overcrowded communal areas. By the end of 1989, some 52,300 families or about 5 per cent of all small-scale farming households had been, or where in the process of being, resettled on about 2.6 million hectares since the programme started soon after independence. Since independence, the large-scale commercial farming sector has been reduced by about 18 per cent to approximately 12.2 million hectares. The majority of these families have been resettled in nucleated village settlements (called settlement Model A) with individual arable land holdings of up to six hectares and communal grazing for between 5 and 20 livestock units, depending on the agro-ecological conditions of the area (Zinyama 1986). The question of land tenure within the resettlement areas still remains unresolved. At present, the settlers are state tenants and their right to remain on a scheme derives from three permits granted them by the government: a permit to reside, another permit to cultivate land, and a third to depasture livestock.

Although the land redistribution programme has fallen far short of the government's original target to resettle 162,000 families by 1985, the amount of land transferred to date is still substantial by any standard. Even the Kenyan land transfer programme, one of the largest in post-independent Africa, is small in comparison (Hazlewood 1985).

Thus, the changes in crop production within the small-scale farming sector in Zimbabwe have occurred without any changes to the system of land tenure in the communal areas where most of the increase has taken place. The Zimbabwean experience during the past decade is therefore contrary to the assertion by Smith (1984: 19) that 'in so far as land reform has been a pre-condition for Green Revolution, it has been so mainly through the release of capitalist enterprise and the influence of the State rather than through the invigoration of peasant enterprise'. In Zimbabwe, the key forces of rural transformation have been state intervention and support of peasant enterprise rather than land reform or the energy of a kulak-like capitalist sector. However, it remains to be seen whether the present communal tenure system is resilient enough to accommodate continued increases in both crop output and population pressures. The growing demands to feed and support a rapidly increasing population may eventually lead to changes in the communal land tenure system in the future.

Adoption of new farming methods and growth of output

The principal food and cash crop grown by black small-scale farmers throughout most of Zimbabwe is maize, although small grains such as sorghum, bulrush millet and finger millet, are fairly widespread in the lower rainfall regions in the south and west of the country. Cotton, and to a lesser extent, sunflowers, are also important as cash crops, although the former is mainly concentrated in a belt stretching from the north-west to the north and north-east. Other minor food and cash crops include groundnuts and pulses. Where investments in the form of purchases of hybrid seed and chemical fertilizers are made, or livestock manure is applied, these are usually for the maize crop, although cotton cultivation also requires the purchase of certified seed from the CMB. For most other crops, farmers use seed selected from the previous year's harvest.

Several important innovations have been adopted by black farmers during the past century, notably the plough, crop rotation, row planting, mono-cropping, contour ridging, land manuring and chemical fertilizers, and hybrid seed varieties. Many of these innovations, some of which were compulsorily imposed upon the black population during the colonial period, were the work of D.E. Alvord, an American missionary–agriculturalist who was appointed by the government in 1926 to establish and head the Department of Native Agriculture until his retirement in 1949.

One of the first innovations at the beginning of the century was the introduction of the ox-drawn single furrow plough which gradually replaced the traditional hand hoe as the principal implement for land preparation. The number of ploughs owned by black farmers throughout the country increased steadily (Southern Rhodesia 1952). Even today, the ox-drawn plough is still the most important farming implement among small-scale farmers. Its greatest impact was that it made it possible for farmers to cultivate larger areas of land than ever before. According to figures compiled from estimates supplied by government district officials, the total area under cultivation by black farmers increased threefold between 1911 and 1951 to 1.14 million hectares (Southern Rhodesia 1952). However, although the cultivation of larger areas of land may have raised total food crop production, the gains were short-lived and do not seem to have brought about a sustained increase in land productivity. With the passage of time, output per unit area and per unit of labour would have declined for at least two reasons: first, the increasing soil exhaustion arising from continuous tillage with poor methods of fertility restoration; and second, the extension of cultivation into increasingly marginal areas because of growing population pressures. Poor soil fertility maintenance was one of the contraints faced by black farmers before independence because they did not have enough livestock to provide them with adequate manure. Furthermore, the use of alternative soil regenerative inputs such as chemical fertilizers or other organic and inorganic materials was very limited. According to Yudelman (1964), less than 3 per cent of black farmers used fertilizers in 1960.

Although livestock numbers within the communal areas have wisen over the years in proportion to the increase in the number of cultivators, average herd sizes per household are low and many families do not have any cattle to provide them with either manure or draught power. In recent years, the problem of inadequate cattle has been compounded by the losses that were incurred during the drought years 1982–5 and 1986–7. Studies in various parts of the country show that communal area farmers have on average between two and four hectares of arable land, of which 50 to 80 per cent is planted with maize annually. It has been estimated that the manure produced per year by one mature beast would maintain the fertility of 0.4 hectares of cropland (Republic of Zimbabwe 1982). Thus, assuming (and this is common practice) that all the manure was applied to the maize crop, one hectare would require a minimum of three mature beasts and at least five head of cattle would be needed for two hectares of maize annually. Yet surveys in Mhondoro and Save North communal areas south-west and south-east of Harare respectively during 1983–4 showed that one-third of the families in the former and one-quarter in the latter area had no cattle (Zinyama 1988c). In Save North, a further one-third had less than five head of cattle of all ages. Thus, over half the households could barely obtain sufficient manure from their cattle. In Buhera communal area to the south of Save North, the average herd size in 1983 was nine but 16 per cent of the households had no cattle (Campbell *et al.* 1989). Stanning (1987) reported that in the Hurungwe, Binga, and Bushy areas in the north and north-west of the country average household herd sizes in 1985 were 7, 4 and 4 respectively and that in all three areas approximately one-third of the households had no cattle. The highest proportion (35 per cent) was in Bushu where another 44 per cent had between 1 and 5 head of cattle. In Mangwende, east of Harare, only 46 per cent of the farmers had cattle in 1982 and of these cattle-owners, only 52 per cent applied cattle manure on their maize plots during the 1981/2 season (Shumba 1984). These data clearly show that many rural households do not have enough cattle to provide them with manure for their arable lands. (The various areas mentioned in this chapter are shown in Figure 7.4).

The provision of institutional credit facilities by the Agricultural Finance Corporation (AFC) and the larger proceeds from crop sales that have become available from increased market participation since 1980 have given the farmers the financial means with which to purchase chemical fertilizers for use in place of, or as a supplement to, inadequate cattle manure in order to raise farm productivity. Until the late 1970s, black farmers were denied access to institutional agricultural credit facilities which were available only to white farmers. In April 1987, the AFC started lending to black farmers in the small-scale commercial farming areas. Its lending facilities were extended to communal area farmers the following, 1979/80, season (Zinyama

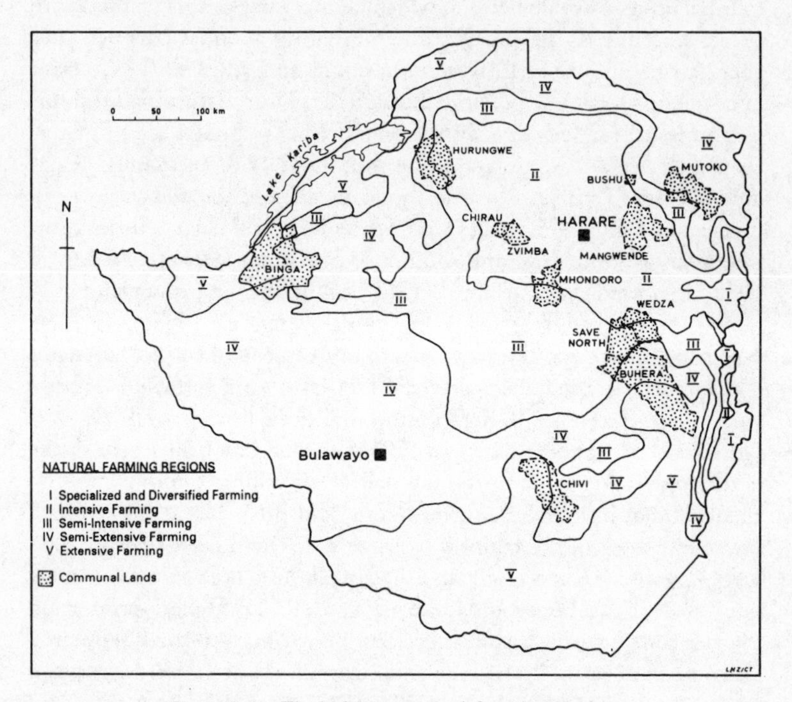

Figure 7.4 Zimbabwe: natural farming regions

1986). The total number of black farmers receiving their crop production inputs (mainly hybrid maize seed, chemical fertilizers and, to a lesser extent, herbicides and pesticides) through short-term or seasonal credit from the AFC increased from 1,290 with a loan value of Z$1.07m for the 1978/9 season to 58,422 valued at Z$46.53m for the 1988/9 season (AFC 1989). The AFC, a parastatal organization, has therefore played a significant developmental role in the diffusion of new farming technologies among the black small-scale farmers during the past decade.

The crucial role of chemical fertilizers in raising crop output and productivity has been demonstrated in an analysis of the technological factors that have been responsible for the increase in maize production within the white large-scale commercial farming sector since the early 1950s (Tattersfield 1982). It was found that the most important factors that have contributed to the increase in maize yield in this sector since 1950 are the application of adequate levels of nitrogenous fertilizers, the use of high-yielding hybrid seed varieties, correct plant population densities, and weed and pest control. Tattersfield calculated that proper utilization of the full package of modern technology arising from research findings would have led to an increase in maize output of 325 per cent over the 1950 yield. Nearly two-thirds (62 per cent) of this increase would have come from the correct application of chemical fertilizers alone, with another 14 per cent from the use of hybrid seed varieties. Thus, the two innovations alone account for three-quarters of the potential increase in maize output since 1950. Yet, until independence, black farmers were denied access to these innovations by a variety of institutional and economic constraints such as the lack of adequate extension services, poor transport, and limited access to marketing and credit facilities. It therefore follows that the removal of these constraints after 1980 and the subsequent widespread adoption of these two techniques would have a major impact on maize production in the small-scale farming areas. The transfer of such yield-increasing technologies from the research stations and the commercial farming sector and their successful adoption by small-scale farmers is therefore a major component to the success of the rural development effort in post-independence Zimbabwe.

Surveys conducted among communal area farmers in various parts of the country show that there has been very rapid adoption of the two yield-increasing innovations – hybrid maize seed varieties and chemical fertilizers – since independence. Short season drought-resistant hybrid seed varieties were first released in the country in the early 1970s. They are better suited to the low and variable rainfall conditions that are characteristic of most communal areas. Initial adoption before 1980, without fertilizers, is estimated to have increased communal area maize output by 30 per cent (Rohrbach 1987). The adoption of hybrid seed varieties has generally been more rapid than that of chemical fertilizers because of the lower costs, relative technical simplicity, and greater cost divisibility of the former. The cost to a farmer of hybrid maize seed for one hectare of cropland is considerably less than the cost of fertilizers, especially if the latter is applied as recommended by agricultural extension staff. For the sandy granitic soils that are widespread in the communal areas, extension staff usually recommend minimum application levels of 250–300 kg of basal and top dressing nitrogenous fertilizers per hectare of maize, with less basal where an adequate amount of livestock manure (four tonnes per hectare per year) has been applied at the start of the rainy season. Another reason for the more rapid adoption of hybrid seed is that it is more readily available to farmers through the numerous small rural retail shops. On the other hand, only the bigger and better capitalized outlets at the larger rural service centres stock chemical fertilizers. Otherwise, farmers have to get their fertilizers from the main urban centres by public bus or private transport or on credit through the AFC. In the latter case the inputs are delivered to a convenient mutually agreed local distribution point.

Sales of fertilizers to the communal farming sector were small and stagnant before independence, with an annual average of 23,300 tonnes during the six-year period 1974–80 (Rohrbach 1987). They tripled during the 1980/1 season, mainly because of massive purchases associated with the government's programmes of post-war rural rehabilitation and reconstruction. Under the programme, rural households were provided at the start of the 1980/1 season with free input packs that included fertilizers, hybrid maize seed, and pesticides. For many farmers, that was the first time they had used fertilizers. By 1985/6, fertilizer sales to the communal sector had increased to an estimated 130,000 tonnes. Sales of hybrid seed have shown a similar increase, from an annual average of 3,540 tonnes between 1974 and 1980 to 9,650 tonnes during the 1980/1 season and an estimated 20,250 tonnes by 1985/6. Thus, between the 1979/80 and 1985/6 seasons, sales of fertilizers and hybrid seed to the communal sector increased by 380 and 370 per cent respectively.

THE CONSEQUENCES OF CHANGING PRODUCTION: SOME CASE STUDIES

This section reviews some of the research findings on the adoption of the two innovations by farmers in a number of communal areas, notably Mangwende, Chivi, Mhondoro, and Save North. Sales of maize from Mhondoro to the GMB, which averaged 1,014 tonnes per year during the pre-independence period 1975–80, quickly increased to 14,038 tonnes following the 1982/3 harvest. There was a drop in sales back to pre-1980 levels during the next two years 1983–5 because of the severe drought. However, deliveries from the area rose again the following year to 15,113 tonnes after the 1985/6 harvest once the drought was over. Sales from Mangwende also rose from less than 10,000 tonnes in 1980 to over 70,000 tonnes in 1986, again with fluctuations around 1983 due to the drought. In Save North, maize deliveries to the GMB have also increased from a pre-independence annual average of 95 tonnes during the period 1975–80 to over 8,000 tonnes by 1986. Maize sales from Chivi have not been as dramatic because the area is in the low rainfall zone which experiences frequent droughts and crop failures.

During 1986 and 1987, Rohrbach (1987) conducted household surveys in Mangwende (a high rainfall area some 80 km east of Harare) and Chivi (a low rainfall area 370 km south of the capital). The research aimed to identify the factors responsible for the large post-independence increases in communal area maize production and market sales. He found that maize yields had increased sharply in both areas in recent years. The increase was attributed initially to an expansion in area planted, but 'larger and more consistent increases, however, resulted from the rapid adoption of improved management practices' (Rohrbach 1987: 169). The most crucial practices were the adoption of short season drought resistant maize seed varieties and chemical fertilizers. Rohrbach found that 42 per cent of the producers in Mangwende and one-half in Chivi had adopted the hybrid varieties by 1975. Five years later, the proportions had risen to 77 per cent and 90 per cent respectively. By 1985, 99 per cent in Mangwende and all the producers in Chivi were using hybrid seed. While much of the yield gain associated with the use of hybrid seed alone was achieved before 1980, its full potential was not exploited until higher rates of fertilizer use were achieved after independence.

In Mangwende, the adoption of fertilizers generally occurred up to two years after hybrid maize seed was first used. By 1975, 45 per cent of the farmers had tried fertilizers, rising to 70 per cent in 1980 and 95 per cent in 1986. Not only did the proportion of farmers using fertilizers increase after independence, but so did application levels per hectare of cropland. By 1986, the average maize yield from fields without fertilizers was only 527 kg per hectare, whereas farmers applying the recommended levels averaged almost three tonnes per hectare. The figures represent approximately 9 per cent and 52 per cent respectively of the average yield per hectare in the large-scale commercial farming sector. Thus, most of the gain in maize yields and sales in Mangwende since independence is attributed to increased use of chemical fertilizers.

In Chivi on the other hand, few farmers had tried chemical fertilizers; in 1980 less than 1 per cent, and by 1986 only 17 per cent. Even those farmers who were using fertilizers were applying such low levels per hectare that the resultant increase in yield was small. One possible explanation not explored by Rohrbach for the low rate of fertilizer adoption in Chivi, in contrast to Mangwende, is the widely held view by farmers in low rainfall areas that the application of fertilizers can 'burn out' a crop should the rains prove inadequate or erratic. The non-adoption of fertilizers is therefore one way of reducing the risk of total crop loss in the event of a drought.

Similar adoption patterns for hybrid seed varieties and chemical fertilizers were observed in two other communal areas, Mhondoro (about 40 km south-west of Harare) and Save North (which is over 150 km to the south-east), during surveys conducted in 1983 and 1984 by Zinyama (1988c). In 1980, 57 per cent of the farmers in Mhondoro and 53 per cent in Save North were using hybrid maize seed. By 1983, levels of adoption had increased to 93 per cent in Mhondoro and 7 per cent in Save North. Rates of adoption of fertilizers increased from 42 per cent in 1980 to 89 per cent in 1983 in Mhondoro and from 23 to 62 per cent respectively in Save North. It was also found that although the adoption of fertilizers had increased rapidly in the post-independence period, many user-households in both areas were not applying the recommended quantities per hectare for optimum yields. In Mhondoro, only 23 per cent of all the households were using the recommended minimum quantity per hectare during the 1982/3 season. In Save North, the proportion of households using less than the recommended minimum was even lower, with greater consequent loss of output. In Mhondoro during the previous season (1981/2), three-quarters of those farmers who used little (under 50 kg) or no fertilizers obtained an average of less than ten bags of maize each (1 bag = 91 kg), and only 5 per cent obtained more than thirty bags each. One-fifth of the households that used 101–300 kg of fertilizers obtained less than ten bags of maize each and 25 per cent obtained more than thirty bags. On the other hand, 82 per cent of the farmers who applied more than 300 kg of fertilizers obtained more than thirty bags, with 11 per cent harvesting over 100 bags each.

These data provide further evidence of the crucial role of fertilizer adoption and its correct application in explaining the increases in maize yields and sales within the small-scale farming sector during the past decade. It is also worth noting that farmers in both Mhondoro and Save North listed the lack of money for purchasing inputs (hybrid seed and chemical fertilizers) as one of their principal constraints against increased crop production (Zinyama 1988d). The provision of credit facilities for input purchasing from the AFC, the use of proceeds from the sale of crops, and the formation of self-help rural savings clubs are all helping to alleviate this constraint.

Similar results on the adoption of chemical fertilizers and hybrid seed varieties and their impact on maize yields and sales have been reported from other communal farming areas of Zimbabwe. For instance, in Wedza some 120 km south-east of Harare, Callear (1984) found that all but two of her sample of 98 households were using hybrid maize seed, whilst 89 used some chemical fertilizers during the 1981/2 season. In Chirau and Zvimba communal areas north-west of Harare, 75 per cent of the farmers were using fertilizers by mid-1982 (Gobbins and Prankerd 1983). Regression analysis showed that the amount of fertilizers used explained 75 per cent of the variation in maize yields by household and their use was viewed by the farmers as a highly profitable investment.

It has been shown in the preceding paragraphs that there has been rapid adoption of the two yield-increasing technologies (hybrid seed varieties and chemical fertilizers) for maize production by small-scale farmers in Zimbabwe since independence. Utilization of these innovations is but one of the factors responsible for the expansion in both total output and productivity that have taken place in the sector during the past decade. It is doubtful whether state intervention and the provision of the various institutional infrastructure and services alone, without the adoption of the two technologies by the farmers themselves, would have achieved the same dramatic results. The expansion both in terms of crop output and in the sector's contribution to national food self-sufficiency would seem to mark the onset of an agricultural revolution, using three of the four criteria specified by Grigg (1984). The exception is that the phenomenal growth has occurred in the absence of any tenurial changes within the communal farming sector. However, as in Asia, the growth currently taking place partly as a result of the adoption of new farming practices is not equitably distributed throughout all regions and among all farming households. Aggregate national and regional data can hide wide village-level disparities in food production and on-farm incomes from crop sales. Some of these subnational spatial and regional disparities, arising mainly from differences in agro-ecological conditions, have been discussed elsewhere (Stanning 1987; Zinyama 1988a). The following section examines the social context of technology adoption and agricultural transformation in the small-scale farming sector in Zimbabwe.

THE SOCIO-ECONOMIC CONTEXT OF AGRICULTURAL TRANSFORMATION

Although a large majority of farmers in the small-scale sector are now using fertilizers and hybrid seed, not all households have benefited significantly from the adoption of these innovations. Crop output, disposable surpluses, and on-farm incomes vary considerably between households, even within the same villages. A large number of families in the communal areas, although they may have tried the two innovations, are none the less unable to produce enough food to meet their household requirements, even after ample rainfall. For instance, Chopak (1989) reported that even in the good season of 1987/8, 59 per cent of the households in Mutoko and Buhera communal areas in the north-east and south-east of the country respectively, produced insufficient food to meet household calorie needs, with the proportion of such food-deficit families rising by another 20 per cent during a poor rainfall season. It is therefore not surprising that malnutrition remains a major public health problem in the rural areas of Zimbabwe, in spite of the large increases in national food production in recent years. Surveys conducted by the Ministry of Health and other health personnel show high incidence of underweight, wasting, and undernourishment among rural children, particularly those from poor households whose families are unable to produce enough food for their domestic use (Sanders and Davies 1988).

This suggests that technology adoption alone does not guarantee adequate or increased food supply for all farming households. Factors contributing towards these differences in output between families include the households' access to and ownership of farming resources such as land, draught cattle, and basic implements; the household demographic structure and the availability of (either family or hired) labour; whether or not the male head of household is permanently resident at home; the kinds of remittances, including purchased farm inputs, and the frequency these are sent home; opportunities for household members to obtain cash for use in purchasing farm inputs from local off-farm employment or other income generating activities; farming experience and educational level of the household's principal decision-maker on agricultural matters; the age and gender of the household head; and participation in local farmer organizations for easier access to credit facilities and technical expertise. The influence of some of these factors on individual household crop production has been examined by a number of researchers (Callear 1984; Gobbins and Prankerd 1983; Shumba 1984; Zinyama 1987). Shumba (1984) has argued that ownership of adequate draught power (ideally a minimum of four strong oxen) for proper land preparation and timely ploughing at the start of the rainy season, is the critical factor influencing crop output in the communal farming areas.

Gobbins and Prankerd (1983) classified farmers in four communal areas north-west of Harare into two categories according to whether or not they had undergone formal on-farm training in agriculture under the supervision of their local government extension worker at the end of which they qualified as 'master farmers'. In all four areas, master farmers had access to more land and family labour, owned more cattle and draught power, used more purchased inputs especially fertilizers, and generally had a more balanced and favourable mix of resources than non-master farmers. As a result, master farmers obtained higher yields and incomes from crop sales. In two of the areas, Chirau and Zvimba, average fertilizer applications for 1981/2 were 9 bags (50 kg per bag) per hectare for master farmers and 3.5 bags per hectare for non-master farmers. Maize output that season was 28 bags per hectare for the former and 19 bags per hectare for the latter. Average earnings per household from sales of all crops were Z$645 for master farmers and Z$233 for non-master farmers. Thus, despite the fact that 75 per cent of all households in Chirau and Zvimba had already adopted fertilizers by that time, there existed wide differences in output, sales, and on-farm incomes between households. These differences were not merely because some families had adopted the new technology while others had not. The overall resource position of the individual household was the critical differentiating factor.

Studies in several parts of the country have identified at least two categories of households that are particularly disadvantaged with regard to ownership of resources and, hence, crop output. The first category comprises female-headed households, that is widows and divorcees rather than women whose husbands are absent in town. These households generally have less arable land, family labour, draught cattle and other assets than other households (Zinyama 1987; Callear 1984). They frequently rely on hired labour or on relatives to plough their lands for them so therefore land preparation and planting are completed later than that of other households. As a result their crops do not benefit from the early rains which are so vital for optimum yields, given the variable seasonal rainfall conditions in Zimbabwe. The second category comprises younger male-headed households. These usually have less arable land than older households. Their smaller land holdings are symptomatic of the growing problems of population pressure and land shortage in the communal areas. Younger, recently established families are frequently compelled to clear the little grazing and other marginal land that still remains, or the parents may subdivide and allocate part of their own lands to their newly married sons. Younger families are also less likely to have accumulated an adequate range of farming assets, including cattle and a plough. They have fewer and younger children to provide labour and their male heads are likely to be away from home in urban low-wage occupations or searching for employment, thereby reducing the amount of family labour available for agriculture. Crop output and food supplies for these households will therefore be low, even though the family may be using hybrid seed varieties and some chemical fertilizers.

CONCLUSION

The attainment of food self-sufficiency is a multi-faceted problem that requires the cooperation of both governments and non-governmental organizations engaged in the provision of the necessary support services on one hand, and the full participation of the smallholder farmers on the other. The experience of Zimbabwe during the past decade is a clear demonstration of what can be achieved through such cooperative effort. This chapter has shown that supportive state policies have facilitated the adoption of yield-increasing technologies by black small-scale farmers. The expansion of crop output and the sector's increased contribution to total agricultural output, national food security, and export earnings point to an incipient agricultural revolution in the small-scale farm sector of Zimbabwe. However, the transformation to date has been confined principally to maize and cotton. In one respect, the emphasis on maize has assured the country of its basic food requirements. None the less, in the long-run, there is need to expand the sector's economic base to include other food and cash crops such as wheat, sunflowers, and soya beans. Unfortunately, such diversification will be costly in terms of capital investment (for instance wheat in Zimbabwe is a dry season crop that will require the development of small-scale irrigation projects for smallholder farmers to venture into it), research and the development of ecologically suitable seed varieties and pest control techniques suited to the needs and management capabilities of small-scale farmers, and more intensive extension support for farmers venturing into higher technology crops. It has also been shown that technology adoption alone does not explain the expansion in crop output of the past decade and the differences in the performance of individual households. Technology adoption is only effective in raising food and cash crop output where the household is adequately endowed with the necessary farming resources such as land, labour, draught power, and implements. Without these, some households will remain at subsistence food production levels even though they may be using some of the yield-increasing technologies. For such disadvantaged households, the benefits from these technologies will therefore not be fully realized.

NOTES

1 The small-scale farming sector in this chapter includes the communal farming areas, the small-scale commercial farming areas (formerly known as the African Purchase Areas before independence) and the post-independence resettlement areas. All three areas are inhabited by black farmers. The large-scale commercial farming sector was, before independence, exclusively for white farmers. However, since 1980, a small number of wealthy blacks have acquired farms in this sector. Much of the increase in crop production and sales within the small-scale farm sector has taken place in the formerly subsistence communal farming areas. The communal areas contribute some four-fifths of the annual maize sales and an even larger proportion of the seed cotton sold by the small-scale farming sector as a whole. The small-scale commercial and resettlement farmers account for more or less equal proportions of the remaining 20 per cent of the maize sold to the Grain Marketing Board each year.

2 Zimbabwe is divided for commercial farming purposes into five agro-ecological zones, commonly called Natural Farming Regions (see Figure 7.4). The division is based primarily on rainfall amount and reliability. Conditions become increasingly marginal for crop production from Natural Region I (over 900 mm per year) to Natural Regions IV and V (under 650 mm) as rainfall decreases in both amount and reliability. Natural Regions IV and V are subject to periodic seasonal droughts and severe dry spells during the rainy season. Natural Farming Regions I–111 are suited for intensive to semi-intensive crop production whereas Natural Regions IV and V should ideally be used for semi-extensive and extensive livestock and game ranching. Land alienation during the colonial period was such that 73 per cent of the land in the agriculturally more favourable Natural Regions I–II was set aside for white settlement. On the other hand, three-quarters of the communal farming areas, compared with

49 per cent of the white-owned farmland, were situated within Natural Regions IV and V. The black farming areas also coincide with the distribution of extensive areas of exposed inselbergs where the granitic parent rocks break down into inherently infertile sandy soil.

3 On all graphs the small-scale sector encompasses the communal, small-scale commercial and resettlement sub-sectors. In the case of cotton seed it also includes output from the AROA estates.

BIBLIOGRAPHY

Agricultural Finance Corporation (annual) *Annual Report Accounts*, Harare.

Agricultural Marketing Authority (annual) *Cotton Situation and Outlook Report*, Harare.

—— (annual) *Grain Situation and Outlook Report*, Harare.

Armstrong, A. (1987) 'Master plans for Dar es Salaam, Tanzania', *Habitat International* 11 (2), 123–45.

Arntzen, J.W. (1989) *Environmental Pressure and Adaption in Rural Botswana*, Amsterdam, Vrije Universiteit.

Arrighi, G. (1973) 'Labour supplies in historical perspective: a study of the proletarianization of the African peasantry in Rhodesia', in G. Arrighi and J.S. Saul (eds), *Essays in the Political Economy of Africa*, New York, Monthly Review Press, 180–234.

Ashcroft, J. (1987) *Regional Services Councils: A Preliminary Assessment*, Cambridge, Jubilee Centre Publications.

Assocom (1987) 'Appeal to the Minister of Finance', Quoted from *The Star*, Johannesburg, 14 May.

Atkinson, A. and Heymans, C. (1988) 'The future: what do the practitioners think?' in C. Heymans and G. Totemeyer (eds), *Government by the People?*, Cape Town, Juta, 143–53.

Bamberger, M., Sanyal, B., and Valverde, N. (1982) *Evaluation of Sites and Services Projects: The Experience from Lusaka, Zambia*, World Bank Staff Working Paper No. 548, Washington, DC.

Batezat, E., Mualo, M., and Truscott, K. (1988) 'Women and independence: the heritage and the struggle', in C. Stoneman (ed.), *Zimbabwe's Prospects*, London, Macmillan, 153–73.

Baylies, C. (1981) 'Imperialism and settler capitalism: friends or foes', *Review of African Political Economy* 18, 116–26.

Bekker, S. (1988) 'Devolution and the state's programme of reform at the local level', in C. Heymans and G. Totemeyer (eds), *Government by the People?*, Cape Town, Juta, 25–33.

—— and Humphries, R. (1985) *From Control to Confusion: The Changing Role of Administration Boards in South Africa*, Pietermaritzburg, Shuter and Shooter.

—— LePere, G., and Tomlinson, M. (1986) 'Dissension over Regional

Services Councils: the stances of opposition groupings to local government reform', *Social Dynamics* 12 (2), 48–63.

Bekure, S. and Dyson-Hudson, N. (1982) 'The operation and viability of the Botswana Second Livestock Development Project (1497-BT): selected issues', Gaborone, Ministry of Agriculture.

Biermann, W. (1988) 'Tanzanian politics under IMF pressure', in M. Hodd (ed.), *Tanzania After Nyerere*, London, Pinter Publishers, 175–83.

——— and Kossler, R. (1981) 'The settler mode of production: the Rhodesian case', *Review of African Political Economy* 18, 106–16.

Blackie, M.J. (1987) 'The elusive peasant: Zimbabwe's agriculture policy, 1965–1986', in M. Rukuni and C.K. Eicher (eds), *Food Security for Southern Africa*, UZ/MSU Food Security Project, Department of Agricultural Economics and Extension, University of Zimbabwe, Harare, 114–44.

Blankhart, S.T. (1986) 'Urban transport in Lusaka', in G.J. Williams (ed.), *Lusaka and its Environs*, Lusaka, Zambia Geographical Association, 259–66.

Botswana, Government of (1973) 'National policy for rural development', Government Paper no. 2, Government Printer, Gaborone.

Brand, V. (1986) 'One-dollar workplaces: a study of informal activities in Magaba, Harare', *Journal of Social Development in Africa*, 1 (2), 52–74.

Bratton, M. and Burgess, S. (1987) 'Afro-Marxism in a market economy: public policy in Zimbabwe', in E.J. Keller and D. Rothchild (eds), *Afro-Marxist Regimes: Ideology and Public Policy*, Boulder, Lynne Rienner, 199–222.

Bryceson, D.F. (1984) 'Urbanization and agrarian development in Tanzania with special reference to secondary cities', Report to the International Institute for Environment and Development.

——— (1985a) 'Women's proletarianisation and the family wage in Tanzania', in H. Asfar (ed.), *Women, Work and Ideology in the Third World*, London, Tavistock, 128–52.

——— (1985b) 'Food and urban purchasing power: the case of Dar es Salaam, Tanzania', *African Affairs* 84 (337), 499–522.

——— (1987) 'A century of food supply in Dar es Salaam', in J.I. Guyer (ed.), *Feeding African Cities*, Manchester, Manchester University Press, 155–202.

Burdette, M. (1988) *Zambia: Between Two Worlds*, Boulder, Westview.

Burgess, R. (1978) 'Petty commodity housing or dweller control? A critique of John Turner's views on housing policy', *World Development* 6, 1105–33.

——— (1982) 'The politics of urban residence in Latin America', *International Journal of Urban and Regional Research* 6, 465–80.

Butcher, C. (1986) *Low Income Housing in Zimbabwe: A Case Study of the Epworth Squatter Upgrading Programme*, Department of Rural and Urban Planning, Occ. paper no. 6, University of Zimbabwe, Harare.

Callear, D. (1984) 'Land and food in the Wedze communal area', *Zimbabwe Agricultural Journal* 81 (4), 163–8.

Cameron, R. (1988) 'The institutional parameters of local government restructuring in South Africa', in C. Heymans and G. Totemeyer (eds), *Government by the People?*, Cape Town, Juta, 49–62.

Campbell, B.M., Du Toit, R.F. and Attwell, C.A.M. (eds) (1989) *The Save Study*, Harare, University of Zimbabwe Publications.

Campbell, J. (1988) 'Tanzania and the World Bank's urban shelter project', *Review of African Political Economy* 41, 5–18.

Chaskalson, M., Jochelson, K., and Seekings, J. (1987) 'Rent boycotts and urban political economy', *South African Review* 4, Johannesburg, Ravan Press, 53–74.

Cheater, A.P. (1979) *The Production and Marketing of Fresh Produce Amongst Blacks in Zimbabwe*, supplement to Zambesia, University of Zimbabwe, Harare.

Cheater, A.P. (1988) 'Contradictions in modelling consciousness: Zimbabwean proletarians in the making?', *Journal of Southern African Studies* 14 (2), 291–303.

Cheatle, M.E. (1986) 'Water supply, sewerage and drainage', in G.J. Williams (ed.), *Lusaka and its Environs*, Lusaka, Zambia Geographical Association, 250–8.

Chigume, S.M. and Shaffer, J.D. (1969) 'Grain marketing by communal farmers in Zimbabwe: preliminary results from Mutoko, Mudzi and Buthera districts', in G.D. Mudimu and R.H. Bernsten (eds), *Household and National Food Security in Southern Africa*, UZ/MSU Food Security Research Project, Department of Agricultural Economics and Extension, University of Zimbabwe, Harare, 223–39.

Chikowore, E. (1989) 'Harare: past, present and future', paper presented at a conference on 'Harare: City in Transition', University of Zimbabwe.

Chitoshi, C.M. (1984) 'Suggestions for improving the financing of local administration in Zambia', *Planning and Administration* 11 (2), 16–22.

Chopak, C.J. (1989) 'Family income sources and food security', in G.D. Mudimu and R.H. Bernsten (eds), *Household and National Food Security in Southern Africa*, UZ/MSU Food Security Research Project, Department of Agricultural Economics and Extension, University of Zimbabwe, Harare, 193–208.

Christopher, A.J. (1983) 'From Flint to Soweto: reflections on the colonial origins of the apartheid city', *Area* 15, 145–9.

—— (1987a) 'Apartheid planning in South Africa: the case of Port Elizabeth', *Geographical Journal* 153, 195–204.

—— (1987b) 'Race and residence in colonial Port Elizabeth', *South African Geographical Journal* 69, 3–20.

City of Harare (1987) 'Highfield local development plan', Department of Works, Planning and Development Division, Harare.

Cobbett, W., Glaser, D., Hindson, D., and Swilling, M. (1985) 'Regionalisation, federalism and the reconstruction of the South African state', *South African Labour Bulletin* 10, 87–116.

Cokorinos, L. (1984) 'The political economy of state and party formation in Zimbabwe', in M. Schatzburg (ed.), *The Political Economy of Zimbabwe*, New York, Praeger, 8–54.

Colclough, C. (1988) 'Zambian adjustment strategy – with and without the IMF', *IDS Bulletin* 19 (1), 51–60.

—— and McCarty, S. (1980) *The Political Economy of Botswana: A Study of Growth and Distribution*, Oxford, Oxford University Press.

—— and Fallon, P. (1983) 'Rural poverty in Botswana: dimensions, causes and constraints', in D. Ghai and S. Radwan (eds), *Agrarian Policies and Rural Poverty in Africa*, Geneva, ILO.

Conyers, D.A. (1983) 'Decentralization: the latest fashion in development administration?', *Public Administration and Develoment* 3, 97–107.

——— (1985) 'Rural regional planning: towards an operational theory', *Progress in Planning* 23.

——— and Hills, P. (1984) *An Introduction to Development Planning in Developing Countries*, Chichester, Wiley.

——— and Warren, D.M. (1988) 'The role of integrated rural development projects in developing local institutional capacity', *Manchester Papers on Development* 4 (1), 28–41.

Cook, G.P. (1986) 'Khayelitsha – policy change or crisis response?', *Transactions, Institute of British Geographers, New Series* 11, 57–66.

Corbett, P. (1982) 'Council housing for low-income Indian families in Durban: objectives, strategies and effects', in D.M. Smith (ed.), *Living Under Apartheid*, London, Allen & Unwin.

Cornia, G.A., Jolly, R., and Stewart, F. (eds) (1988) *Adjustment with a Human Face*, Oxford, Clarendon.

CSO (1986) *1985/86 Labour Force Survey*, Government Printer, Gaborone.

——— (1987) *Population Projections 1981–2011: 1981 Population and Housing Census* 5, Government Printer, Gaborone.

Davies, R.J. (1972) 'Changing residential structures in South African cities 1950–70', in P. Adams and F.K. Helleiner (eds), *International Geography*, Toronto, Toronto University Press, 21–49.

——— (1981) 'The spatial formation of the South African city', *Geo Journal: Supplementary Issue* 2, 59–72.

——— (1986) 'When! – reform and change in the South African city', *South African Geographical Journal* 68, 3–17.

Davies, R. (1988) 'The transition to socialism in Zimbabwe: some areas for debate', in C. Stoneman (ed.), *Zimbabwe's Prospects*, London, Macmillan, 18–31.

Davies, W.J. (1971) *Patterns of Non-White Population Distribution in Port Elizabeth with Special Reference to the Application of the Group Areas Act*, University of Port Elizabeth, Institute of Planning Research, Special Publication Series B, no. 1, Port Elizabeth.

Davis, B. and Dopcke, W. (1987) 'Survival and accumulation in Gutu: class formation and the rise of the state in colonial Zimbabwe, 1900–1939', *Journal of Southern African Studies* 14 (1), 64–98.

De Coning, C., Fick, J., and Olivier, N. (1986) *Residential Settlement Patterns: A Pilot Study of Socio-Political Perceptions in Grey Areas of Johannesburg*, Department of Development Studies, Rand Afrikaans University.

Despande, V.W. and Masebu, H. (1986) 'Land for housing Tanazania's urban poor', in K.P. Bhattacharya (ed.), *Human Settlements in Developing Countries: Appropriate Approach and Alternatives*, Calcutta, Centre for Human Settlements, 77–87.

De Tolly, J. (1986) *Regional Services Councils: An Explanation: An Appraisal*, Johannesburg, Black Sash.

Drakakis-Smith, D.W. (1985) 'The changing economic role of women in urbanization: a preliminary analysis from Harare', *International Migration Review* 18 (4), 1277–92.

——— (1990) 'Food for thought or thought about food: urban food distribution systems in the Third World', in R. Potter and A. Salau (eds), *Cities and Development in the Third World*, London, Mansell, 100–21.

——— and Kivell, P.T. (1990) 'Food production, purchasing and exchange in Harare, Zimbabwe', in A. Findlay, R. Paddison, and J. Dawson (eds), *Urban Retailing in the Third World*, London, Routledge, 156–85.

Dusseldorp, D.B.M.W. (1980) 'The process of planned development', in D.B.W.M. Dusseldorp and J.M. van Staveren (eds), *Framework for Regional Planning in Developing Countries*, Wageningen, ICRI.

Egner, E.B. (1978) 'District Development in Botswana', Report to SIDA, Gaborone.

——— (1987) 'The District Councils and Decentralization 1978–1987', Report to SIDA, Gaborone.

Evans, S. (1985) *Local Government: the Reform Process*, Conference of the South African Institute of Town and Regional Planners, Durban.

Evers, H.D. (1984) 'Urban landownership, ethnicity and class in Southeast Asian cities', *International Journal of Urban and Regional Research* 8, 481–96.

Fair, D. (1984) 'Tanzania: some aspects of urban growth and policy', *Africa Insight* 14 (1), 33–40.

Farmer, B.H. (1981) 'The "Green Revolution" in South Asia', *Geography* 66 (3), 202–7.

Fimbo, G.M. (1988) 'Planned urban development versus customary law in Tanzania', paper presented to the Commonwealth Association of Planners, Africa Regional Conference on Urbanization Strategies for SADCC Countries, Ardhi Institute, Dar es Salaam.

Floyd, B.N. (1959) 'Changing patterns of African land use in Southern Rhodesia', *The Rhodes–Livingstone Journal* 25, 20–39.

Food and Agricultural Organization (FAO) (1986) *Policy Options for Agrarian Reform in Zimbabwe: A Technical Appraisal*, paper submitted by FAO for the consideration of the Government of Zimbabwe, Harare.

Garside, J. (1987) ' "Open" areas in Cape Town – the struggle for their identity', Regional Topic Paper no. 87/8, South African Institute of Race Relations, Cape Western Region.

Gasper, D. (1987) 'Development planning and decentralization in Botswana', mimeo.

——— (1988) 'Rural growth points and rural industries in Zimbabwe: ideologies and policies', *Development and Change* 19, 425–66.

Giliomee, H. and Schlemmer, L. (eds) (1985) *Up Against the Fences: Poverty, Passes and Privilege in South Africa*, Cape Town, David Philip.

Gobbins, K.E. and Prankerd, H.A. (1983) 'Communal agriculture: a study from Mashonaland West', *Zimbabwe Agriculture Journal* 80 (4), 151–8.

Granberg, P. and Parkinson, J.R. (eds) (1988) *Botswana: Country Study and Norwegian Aid Review*, Norway, Chr. Michelsen Institute.

Grigg, D.B. (1984) 'The agricultural revolution in Western Europe', in T.P. Bayliss-Smith and S. Wanmali (eds), *Understanding the Green Revolutions: Agrarian Change and Deveent Planning in South Asia*, Cambridge, Cambridge University Press, 1–17.

Hart, T. and Hardie, G.J. (1987) 'State-sanctioned self-help and self-help homebuilders in South Africa', *Environment and Behaviour* 19, 353–70.

Harvey, S. (1987a) 'Black residential mobility in a post-independence Zimbabwean city', in G.J. Williams and A.P. Wood (eds), *Geographical Perspectives on Development in Southern Africa*, Townsville, Commonwealth Geographical Bureau.

—— (1987b) 'Post-colonial urban change in Zimbabwe – a case study', paper presented at the Commonwealth Geographical Bureau Workshop on Urbanization in Developing Countries, Delhi.

Hayuma, A.M. (1979) 'A review and assessment of the contribution of international and bilateral aid to urban development policies in Tanzania', *Ekistics* 46 (279), 349–61.

—— (1986) 'The reforms and operations of urban local government authorities in Tanzania 1899–1986', Working Research Paper no. 2, Department of Urban and Rural Planning, Ardhi Institute, Dar es Salaam.

Hazlewood, A. (1985) 'Kenyan land-transfer programmes and their relevance for Zimbabwe', *Journal of Modern African Studies* 23 (3), 445–61.

Hendler, P. (1987) 'Capital accumulation and conurbation; rethinking the the social geography of the ''Black'' townships', *South African Geographical Journal* 69, 60–85.

—— Mabin, A. and Parnell, S. (1986) 'Rethinking housing questions in South Africa', *South African Review* 3, Johannesburg, Ravan, 195–207.

Hendler, P. and Parnell, S. (1987) 'Land and finance under the new housing dispensation', *South African Review* 4, Johannesburg, Ravan, 423–32.

Heymans, C. (1988) 'The political and constitutional context of local government restructuring', in C. Heymans and G. Totemeyer (eds), *Government by the People?*, Cape Town, Juta, 34–48.

Hinderink, J. and Sterkenburg, J.J. (1987) *Agricultural Commercialisation and Government Policy in Africa*, London, Routledge.

Holm, J.D. (1982) Liberal democracy and rural development in Botswana, *African Studies Review* 25, 83–102.

—— (1985) 'The state, social class and rural development in Botswana', in L.A. Picard (ed.), *The Evolution of Modern Botswana*, London, Rex Collings.

Hodder-Williams, R. (1982) 'A small town in crisis: Marandellas, Zimbabwe', paper presented to the African Studies Association of the UK, Birmingham.

Hoof, P.J.M. van and Jansen, R. (1989a) 'Regional development planning for rural development in Botswana', paper presented to the British Dutch symposium on 'Appropriate Regional Development Strategies', RHBNC, London.

—— (1989b) 'North-east district CFDA: inventory for land use planning', vol. 3, Government Employment Programmes, Gaborone.

—— and Maas, H. Van Der (1989) 'Land use, settlements and the rural poor: wither rural development? A case study from Botswana', paper presented to the IGU conference on 'Limits to Rural Land Use'.

Horn, N. (1986) 'The informal fruit and vegetable market in Greater Harare', Working Paper 4/86, Department of Land Management, University of Zimbabwe.

Horrell, N.W. (1981) 'The potential growth of urban squatter settlements in the capital city of Zimbabwe', mimeo, Department of Housing Services, Harare.

Hudson, P. and Sarakinsky, M. (1986) 'The case of the urban African bourgeoisie', *South African Review 3*, Johannesburg, Ravan, 169–85.

Humphries, R. (1988) 'Intermediate state responses to the black local authority legitimacy crisis', in C. Heymans and G. Totemeyer (eds), *Government by the People?*, Cape Town, Juta, 105–18.

Hyden, G. (1983) *No Shortcuts to Progress: African Development*, Management in Perspective, London, Heinemann.

International Labour Office (1988) *Distributional Aspects of Stabilisation Programmes in the United Republic of Tanzania 1979–1984*, Geneva, ILO.

Jaeger, D. and Huckaby, J.D. (1986) 'The garden city of Lusaka: urban agriculture', in G.F. Williams (ed.), *Lusaka and its Environs*, Lusaka, Geographical Association of Zambia, 267–77.

Jones, N. and O'Donnell, P. (1980) 'Metropolitan areas', in D. Rowat (ed.), *International Handbook on Local Government Reorganisation: Contemporary Developments*, London, Aldwych Press.

Jordan, J.D. (1984) *Local Government in Zimbabwe*, Gweru, Mambo Press.

Kadhani, X.M. (1986) 'The economy: issues, problems and prospects', in I. Mandaza (ed.), *Zimbabwe: The Political Economy of Transition*, Dakar, Codesira, 99–122.

Kaitilla, S. (1987) 'The land delivery mechanism in Tanzania', *Habitat International* 11 (3), 55–9.

Kalabamu, F.T. (n.d.) 'Problems of a bureaucratic housing land alienation process – the case of urban Tanzania', mimeo.

Kalapula, E.S. (1987) 'Electrification of peri-urban areas in Lusaka', *Geography* 72 (3), 243–5.

Kamulali, T.W.P. (1985) 'Sites and services in Tanzania: a case study in Nyakato, Mwanza', in P. Crooke (ed.), *Management of Sites and Services and Squatter Upgrading Housing Areas*, Occ. Paper 3, Centre for Housing Studies, Ardhi Institute, Dar es Salaam, 43–7.

Kasongo, B.A. (1987) 'Development strategies for the homeless shanty town dwellers: the case of the city of Kitwe, Zambia', *African Urban Quarterly* 2 (3), 298–310.

Kay, G. (1976) 'Population problems and development strategy in Rhodesia', *Scottish Geographical Magazine* 92 (3), 14–60.

Kaynak, E. (1981) 'Food distribution systems: evolution in Latin America and the Middle East', *Food Policy*, May, 79–90.

Kironde, J.M.L. (1988) 'Providing land for development: some lessons from the Tanzania experience', paper presented to the Commonwealth Association of Planners Africa Regional Conference on Urbanization Strategies for SADCC Countries, Ardhi Institute, Dar es Salaam.

Krige, D.S. (1988) *Die Transformasie van die Suid-Afrikaanse Stad*, Research Publication 10, Department of Urban and Regional Planning, University of the Orange Free State, Bloemfontein.

Kulaba, S.M. (1981) *Housing, Socialism and National Development in Tanzania: A Policy Framework*, Centre for Housing Studies, Ardhi Institute, Dar es Salaam.

——— (1985) 'Managing rapid urban growth through sites and services and squatter upgrading in Tanzania: lessons of experience', in P. Crooke (ed.), *Management of Sites and Services and Squatter Upgrading Housing Areas*, Occ. Paper 3, Centre for Housing Studies, Ardhi Institute, Dar es Salaam, 31–42.

—— (1989) 'Local government and the management of urban services in Tanzania', in R.E. Stren, and R.R. White (eds), *African Cities in Crisis: Managing Rapid Urban Growth*, Boulder, Westview, 203–45.

Kuper, L., Watts, H., and Davies, R.J. (1958) *Durban: a Study in Racial Ecology*, London, Jonathan Cape.

Lea, J.P. (1982) 'Government dispensation, capitalist imperative or liberal philanthropy? Responses to the Black housing crisis in South Africa', in D.M. Smith (ed.), *Living under Apartheid*, London, Allen & Unwin, 198–216.

Lear, E. and Winter, C. (1985) 'The new Regional Services Councils Bill: what are its implications for the man in the street?', *Paper No. 85/6*, Johannesburg, South African Institute of Race Relations.

Lemon, A. (1982) 'Migrant labour and "frontier commuters": reorganising South Africa's black labour supply', in D.M. Smith (ed.), *Living under Apartheid*, London, Allen & Unwin, 64–89.

—— (1985) 'The Indian and coloured elections: co-optation rejected', *South Africa International* 15, 84–107.

—— (1987a) 'The South African city after apartheid', paper presented to the Association of American Geographers' Conference, Portland, Oregon.

—— (1987b) *Apartheid in Transition*, Aldershot, Gower.

Leiman, A. (1985) 'Formal/informal sector articulation in the Zimbabwean economy', *Journal of Contemporary African Studies* 4 (1/2), 119–37.

Leon, T. (1987) 'Regional Services Councils', *Businessman's Law* 18, 176–9.

Leontidou, L. (1985) 'Urban land rights and working-class consciousness in peripheral societies', *International Journal of Urban and Regional Research* 9, 533–56.

Loewenson, R. (1988) 'Labour insecurity and health: an epidemiological study in Zimbabwe', *Soc. Sci. Med.* 27 (7), 733–41.

Loewenson, R. and Sanders, D. (1988) 'The political economy of health and nutrition', in C. Stoneman (ed.), *Zimbabwe's Prospects*, London, Macmillan, 133–52.

Lungu, S.F. (1986) 'Mission impossible: integrating central and local administration in Zambia', *Planning and Administration* 13 (1), 16–22.

Luning, H.A. (1981) 'The need for regionalized agricultural development planning: experiences from Western Visayas, Philippines', SEARCA, Laguna.

Maasdorp, G. (1982) 'Informal housing and informal employment: case studies in the Durban metropolitan area', in D.M. Smith (ed.), *Living under Apartheid*, London, Allen & Unwin.

Mabin, A. (1986) Labour, capital, class struggle and the origins of residential segregation in Kimberley, 1880–1920', *Journal of Historical Geography* 12, 4–26.

—— and Parnell, S. (1983) 'Recommodification and working class home ownership: new directions for South African cities', *South African Geographical Journal* 65, 148–66.

McCarthy, J.J. (1987) 'A brief reply', *South African Geographical Journal* 69, 86–8.

McLeod, S. and McGee, T.G. (1990) 'Internationalisation of the food supply system in Hong Kong', in D.W. Drakakis-Smith (ed.), *Economic Growth and Urbanization in Developing Countries*, London, Routledge.

Maddick, H. (1963) *Democracy, Decentralization and Development*, Bombay, Asia.

Maembe, E. and Tomecko, I. (1987) 'Economic promotion in an integrated upgrading project: the case of Kalingalinga Township, Lusaka', mimeo, Institute for African Studies, Lusaka.

Magembe, E.A. (1985) 'Towards effective management and consolidation of squatter upgrading areas: the case of Mwanjelwa settlement, Mbeya, Tanzania', in P. Crooke (ed.), *Management of Sites and Services and Squatter Upgrading Housing Areas*, Occ. Paper 3, Centre for Housing Studies, Ardhi Institute, Dar es Salaam, 48–59.

Magembe, E.A. and Rodell, M.J. (1983) *Housing Production in Selected Areas of Dar es Salaam, Tanzania*, Institute for Housing Studies (BIE), Institute for Housing Studies, Rotterdam.

Martin, R. (1974) 'The architecture of underdevelopment or the route to self determination in design?', *Architectural Design* 10, 626–34.

Materu, J. (1985) *Tanzania Sites and Services: a Case Study for Implementation*, TRP 56 Department of Town and Regional Planning, University of Sheffield.

Mather, C. (1987) 'Residential segregation and Johannesburg's "locations in the sky" ', *South African Geographical Journal* 69, 119–28.

Mawhood, P. (1983) 'The search for participation in Tanzania', in P. Mawhood (ed.), *Local Government in the Third World: The Experience of Tropical Africa*, Chichester, Wiley, 75–105.

Mazambani, D. (1980) 'Woodfuel trade and consumption patterns in Salisbury's townships', *Geographical Association of Zimbabwe* 13, 21–35.

——— (1982a) 'Peri-urban cultivation within Greater Harare', *The Zimbabwe Science News* 16 (6), 134–8.

——— (1982b) 'Exploitation of trees around Harare', *The Zimbabwe Science News* 16 (11), 253–6.

Merrington, G.L. (1981) 'A new look at the low-income housing plan', paper presented to the Local Government Association Annual Conference, Victoria Falls.

Mghweno, J. (1984) 'Tanzania's surveyed plots programme', in G. Payne (ed.), *Low-income Housing in the Developing World*, Chichester, Wiley, 109–23.

Ministry of Finance and Development Planning (MFDP) (1985) 'National Development Plan 6, 1985–91', Gaborone.

——— (1988) 'Evaluation of the Financial Assistance Policy (FAP) and its role in Botswana Business Development', Gaborone.

Ministry of Local Government 'Rural and Urban Development (1987) *Report on the National Symposium on Agrarian Reform in Zimbabwe, Nyanga, 19–23 October 1987*, Department of Rural Development and FAO, Harare.

Morton, F. and Ramsay, J. (eds) (1987) *The Birth of Botswana: A History of the Bechuanaland Protectorate from 1910 to 1966*, London, Longman.

Moyo, N.P. (1988) 'The state, planning and labour: towards transforming the colonial labour process in Zimbabwe', *Journal of Development Studies* 24 (4), 203–17.

Mtiro, I.J. (1988) 'Urban development and planning in Tanzania', paper presented to the Commonwealth Association of Planners Africa Regional Conference on Urbanization Strategies for SADCC Countries, Ardhi Institute, Dar es Salaam.

Musekiwa, E. (1989) 'Low income housing development in Harare: an historical perspective', paper presented at a conference on 'Harare: City in Transition', University of Zimbabwe.

Mutizwa-Mangiza, D. (1986) 'Urban centres in Zimbabwe: intercensal changes, 1969–1982', *Geography* 71 (2), 148–50.

Mwali, M. (1988) 'African city bus operation: a study of the performance of bus operations in Sub-Saharan African cities, with special reference to Lusaka, Dar es Salaam and Nairobi', doctoral dissertation, University College, Cardiff.

Nangati, F. (1982) 'Harare Musika: an urban squatter settlement', mimeo, Department of Social Services, Harare.

Ndhlovu, T.P. (1983) *Industrialization of Zimbabwe: A Test of the Frankian Thesis*, Development Studies Occasional Paper no. 21, University of East Anglia, Norwich.

Nientied, P. and Van der Linden, J. (1985) 'Approaches to low-income housing in the Third World: some comments', *International Journal of Urban and Regional Research* 9, 311–29.

Nilsson, P. (1986) 'Wood: the other energy crisis', in J. Boesen, K.J. Harnevik, J. Koponen, and R. Odgard (eds), *Tanzania: Crisis and Struggle for Survival*, Uppsala, Scandinavian Institute of African Studies, 159–72.

Nnkya, T.J. and Kombe, W. (1986) *Lindi Master Plan, Main Report*, Urban Planning Division, Ministry of Lands, Housing and Urban Development, Dar es Salaam.

Noppen, D. (1982) *Consultation and Non-commitment: Planning with the People in Botswana*, Research Report, no. 13, African Studies Centre, Leiden.

Nyawo, C. and Rich, A. (1981) 'Zimbabwe after independence', *Review of African Political Economy* 18, 89–93.

O'Connor, A. (1988) 'The rate of urbanization in Tanzania in the 1970s', in M. Hood (ed.), *Tanzania After Nyerere*, London, Pinter Publishers, 136–42.

Olivier, N.J.J. (1987a) 'A vehicle for constitutional reform? – another view', in South African Institute of Race Relations, Councils and Controversy: South Africa's New Regional Services Councils, Johannesburg, 27–30.

Paddison, R. (1988) *Ideology and Urban Primacy in Tanzania*, DP35, Centre for Urban and Regional Research, University of Glasgow.

Parnell, S. (1986) 'From Mafeking to Mafikeng: the transformation of a South African town', *Geo Journal* 12, 203–20.

Patel, D. (1984) 'Housing the urban poor in the socialist transformation of Zimbabwe', in M. Schatzburg (ed.), *The Political Economy of Zimbabwe*, New York, Praeger, 182–96.

Phiri, M. (1990) 'The most densely populated suburbs', *Sunday Mail*, 17 June, 11.

Picard, L.A. (1979) 'District councils in Botswana: a remnant of local autonomy', *The Journal of Modern African Studies* 71, 285–308.

—— (1987) *The Politics of Development in Botswana: a Model for Success?*, Boulder, Lynne Reinner.

—— and Morgan, E.P. (1985) 'Policy, implementation and local institutions in Botswana', in L.A. Picard (ed.), *The Evolution of Modern Botswana*, London, Rex Collings, 39–52.

Pirie, G.H. (1984) 'Race zoning in South Africa: board, court, parliament, public, *Political Geography Quarterly* 3, 207–21.

—— (1986) 'More of a blush than a rash: changes in urban race zoning', *South African Review* 3, Johannesburg, Ravan, 186–95.

—— (1987) 'Deconsecrating the holy cow: reforming the Group Areas Act', *South African Review* 4, Johannesburg, Ravan, 402–11.

—— and Hart, D. (1985) 'The transformation of Johannesburg's Black western areas', *Journal of Urban History* 11, 387–410.

Platzky, L. and Walker, C. (1985) *The Surplus People: Forced Removals in South Africa*, Johannesburg, Ravan.

Porteous, D. (1987) 'Regional Services Councils (RSCs): and economic perspctive', Appendix to J. Ashcroft, *Regional Services Comments: A Preliminary Assessment*, Cambridge, Jubilee Centre Publications.

Porter, D.O. and Olsen, E.A. (1976) 'Some critical issues in Government centralization and decentralization', *Public Administration Review* 1, 72–84.

Potts, D. (1987) 'Recent rural–urban migrants to Harare, Zimbabwe: the maintenance of rural ties', paper presented at a Workshop on Rural–Urban Links, Centre for African Studies, SOAS, University of London.

—— and Mutambirwa, C.C. (1989) 'Rural–urban migration in Zimbabwe: an attempt to identify some policy issues', *Proceedings, Geog. Assoc. of Zimbabwe*, 22, 1–8.

Rakodi, C. (1985) 'Self-reliance or survival? Food production in African cities with special reference to Zimbabwe', *African Urban Studies* 21, 53–63.

—— (1986) 'Housing in Lusaka: policies and progress in G.J. Williams (ed.), *Lusaka and Its Environs*, Lusaka, Zambia Geografica Association, 189–209.

—— (1987a) 'Urban plan preparation in Lusaka', *Habitat International* 11 (4), 85–111.

—— (1987b) 'Land, layouts and infrastructure in squatter upgrading: the case of Lusaka', *Cities* 4 (4), 348–70.

—— (1988a) 'Urban planning in Zambia: an assessment of plan preparation and the development control system, with particular reference to Lusaka urban and rural districts', paper presented to a conference on Town Planning in Africa, African Studies Association of the UK and Development Planning Unit, London.

—— (1988b) 'The local state and urban local government in Zambia', *Public Administration and Development* 8, 27–46.

—— (1988c) 'Upgrading in Chawama, Lusaka displacement or differentiation', *Urban Studies* 25, 297–318.

—— (1988c) 'Urban agriculture: research questions and Zambian evidence', *Journal of Modern African Studies* 26 (3), 495–516.

—— (1989) 'The production of housing in Harare, Zimbabwe: components, constraints and policy outcomes', *Trialog* 20, 7–13.

——— (1990) 'Housing production and housing policy in Harare, Zimbabwe', *Journal of Urban Affairs* 12, 1, 135–56.

——— and Mutizwa-Mangiza, N. (1989) *Housing Policy, Production and Consumption in Harare, Zimbabwe: A Review*, Teaching Paper no. 3, Department of Rural and Urban Planning, University of Zimbabwe.

Reilly, W. (1981) 'District development planning in Botswana: studies in decentralization', *Manchester Papers on Development*, no. 3, 128–41.

——— (1983) 'Decentralization in Botswana: myth or reality?', in P. Mawhood (ed.), *Local Government in the Third World*, London, Wiley, 44–56.

Republic of Zimbabwe (1982) *Report of the Committee of Inquiry into the Agricultural Industry*, Chavunduka Commission Report, Harare.

——— (1986) *First Five-Year National Development Plan 1986–1990*, vol. 1, Harare.

——— (1987) *Parliamentary Debates: House of Assembly*, 19 August 1987, 793–4.

Riddell, R. (1981) *Report of the Commission of Inquiry into Prices, Incomes and Conditions of Service*, Salisbury, Government of Zimbabwe.

Robinson, A. (1986) 'Botha aims to modernise apartheid not kill it', *Financial Times*, 6 February.

Rohrbach, D.D. (1987) 'A preliminary assessment of factors underlying the growth of communal maize production in Zimbabwe', in M. Rukuni and C.K. Eicher (eds), *Food Security for Southern Africa*, UZ/MSU Food Security Project, Department of Agricultural Economics and Extension, University of Zimbabwe, Harare, 145–84.

Rondinelli, D.A. (1981) 'Government decentralization in comparative perspective: theory and practice in developing countries', *International Review of Administrative Sciences* 47, 133–45.

——— and Cheema, G.S. (1983) 'Implementing decentralization policies: an introduction', in G.S. Cheema and D.A. Rondinelli (eds), *Decentralization and Development: Policy Implementation in Developing Countries*, Beverly Hills, Sage.

——— McCullough, J.S., and Johnson, R.W. (1989) 'Analysing decentralization policies in developing countries: a political-economy framework', *Development and Change* 20, 57–87.

Rule, S.P. (1988) 'The emergence of racially mixed residential suburbs in Johannesburg, South Africa', paper presented to the 26th IGU Congress, Sydney.

Sachikonye, L.M. (1986) 'State, capital and trade unions', in I. Mandaza (ed.), *Zimbabwe: The Political Economy of Transition*, Dakar, Codesira, 243–74.

Samoff, J. (1979) 'The bureaucracy and the bourgeoisie: decentralization and class structure in Tanzania', *Comparative Studies in Society and History* 21 (1), 30–62.

Sanders, D. and Davies, R. (1988) 'The economy, the health sector and child health in Zimbabwe since independence', *Soc. Sci. Med.*, 27 (7), 723–31.

Sandford, S. (1980) 'Keeping an eye on the TLGP', *NIR Working Paper*, no. 31, Gaborone.

Sanyal, B. (1987a) 'Problems of cost-recovery in development projects:

experience of the Lusaka squatter upgrading and site/service project', *Urban Studies* 24 (4), 285–95.
—— (1987b) 'Urban cultivation amidst modernization: how should we interpret it?', *Journal of Planning Education and Research* 6, 197–207.
Scharzburg, M. (ed.) (1984) *The Political Economy of Zimbabwe*, New York, Praeger.
Schlyter, A. (1987a) 'Commercialisation of housing in upgraded squatter areas: the case of George, Lusaka, Zambia', *African Urban Quarterly* 2 (3), 287–97.
—— (1987b) 'Women householders in Harare', paper presented to the IYSH seminar on Women and Shelter, Harare.
—— (1988) *Women Householders and Housing Strategies: the Case of George, Lusaka*, Gavle, The National Swedish Institute for Building Research.
Schmetzer, H. (1987) 'Slum upgrading and sites and services schemes under different political circumstances: experience from East Africa', *African Urban Quarterly* 2 (3), 276–85.
Schuster, I. (1982) 'Marginal lives: conflict and contradiction in the position of female traders in Lusaka, Zambia', in E. Bay (ed.), *Women and Work in Africa*, Boulder, Westview, 105–26.
Scott, E.P. (1985) 'Lusaka's informal sector in national economic development', *Journal of Developing Areas* 20, 71–100.
Seegers, A. (1988) 'Extending the security network to the local level', in C. Heymans and G. Totemeyer (eds), *Government by the People?*, Cape Town, Juta, 119–39.
Sheriff, F. (1985) 'Housing policies and strategies in Tanzania', *Trialog* 6, 8–14.
Shopo, T.D.C. (1986) 'The political economy of hunger', in I. Mandaza (ed.), *Zimbabwe: The Political Economy in Transition 1980–1986*, Dakar, Codesira, 223–41.
Shumba, E.M. (1984) 'Reduced tillage in the communal areas', *Zimbabwe Agricultural Journal* 81, 235–9.
Sibanda, A. (1988) 'The political situation', in C. Stoneman (ed.), *Zimbabwe's Prospects*, London, Macmillan, 257–83.
Siebolds, P. and Steinberg, F. (1982) 'Tanzania: sites and services', *Habitat International* 6 (1/2), 109–30.
Sijaona, S.T. (1987) 'Women's income generating activities in Epworth, Harare', diploma project, Department of Rural and Urban Planning, University of Zimbabwe, Harare.
Silk, A. (1981) *A Shantytown in South Africa: the Story of Modderdam*, Johannesburg, Ravan.
Simon, B. (1986) 'The not-so-pampered whites', *Financial Times*, 4 August.
Simon, D. (1984) 'The apartheid city', *Area* 16, 60–2.
—— (1985a) 'Contemporary decolonization and comparative urban research in southern Africa', *Comparative Urban Research* 10, 32–41.
—— (1985b) 'Decolonization and local government in Namibia: the neo-apartheid experiment 1977–83', *Journal of Modern African Studies* 23, 507–26.
—— (1986a) 'Desegregation in Namibia: the demise of urban apartheid?', *Geoforum* 17, 289–307.

——— (1986b) 'Regional inequality, migration and development: the case of Zimbabwe', *Tijdschrift v. Econ. en Soc. Geog.* 77 (1), 7–17.

——— (1988) 'Urban squatting, low income housing and politics in Namibia on the eve of independence', in R.A. Obudho and C.C. Mhlanga (eds), *Slum and Squatter Settlements in Sub-Saharan Africa: Towards a Planning Strategy*, New York, Praeger, 245–60.

——— (1989) 'Colonial cities, postcolonial Africa and the world economy: a reinterpretation', *International Journal of Urban and Regional Research* 13, 46–69.

Situma, I.W. (1987) 'Problems of public transportation in Zimbabwe', *African Urban Quarterly* 2 (1), 49–54.

Slater, D. (1989) 'Territorial power and the peripheral state: the issue of decentralization', *Development and Change* 20, 501–31.

Smith, B.C. (1985) *Decentralization. The Territorial Dimension of the State*, London, Allen & Unwin.

Smith, C.T. (1984) 'Land reform as a pre-condition for Green Revolution in Latin America', in T.P. Bayliss-Smith and S. Wanmali (eds), *Understanding Green Revolutions: Agrarian Change and Development Planning in South Asia*, Cambridge, Cambridge University Press, 18–36.

Smith, D.M. (ed.) (1982) *Living Under Apartheid*, London, Allen & Unwin.

——— (1987) 'Conflict in South African cities', *Geography* 72, 153–8.

Smith, J. (1987) 'The transport and marketing of horticultural produce by communal farmers into Harare, Zimbabwe', paper presented at a Workshop on Rural–Urban Links, Centre for African Studies, SOAS, University of London.

Smout, M. (1974) 'Commercial growth and consumer behaviour in suburban Salisbury, Rhodesia', Occasional Paper No. 1, Social Studies Series, University of Rhodesia.

Solomon, D. (1988) 'The financial and fiscal aspects of local government restructuring', in C. Heymans and G. Totemeyer (eds), *Government by the People?*, Cape Town, Juta, 77–94.

South Africa (1976) *Report of the Commission of Inquiry into Matters Relating to the Coloured Population Group*, Pretoria.

——— (1978a) *First and Second Interim Reports of the Commission of Inquiry into the Establishment of Local Authorities in Coloured Group Areas*, Pretoria.

——— (1978b) *Final Report of the Working Committee that Inquired into the Powers, Duties and Functions of Management Committees*, Pretoria.

——— *Report of the Commission of Inquiry into the Establishment of Local Authorities in Indian Group Areas*, Pretoria.

——— (1980) *Report of the Inquiry into the Finances of Local Government Authorities in South Africa*, Pretoria.

——— (1982) *Report of the Working Group on the Report of the Commission of Inquiry into the Finances of Local Government Authorities in South Africa*, Department of Finance, Pretoria.

South African Institute of Race Relations (SAIRR) (1987a) *Councils and Controversy: South Africa's New Regional Services Councils*, Johannesburg.

——— (1987b) *Quarterly Countdown* 5, 6, 7 and 8, Johannesburg.

——— (1987c) *Social and Economic Update* 1, 2, 3 and 4, Johannesburg.

——— (1988a) *Quarterly Countdown* 9, Johannesburg.

——— (1988b) *Social and Economic Update* 6, Johannesburg.

—— (1989a) *Quarterly Countdown* 12, 13, Johannesburg.

—— (1989b) *Social and Economic Update* 7, 8, Johannesburg.

Southall, R.J. (1987) 'Post-apartheid South Africa: constraints on socialism', *Journal of Modern African Studies* 25, 345–74.

Southern Rhodesia (1952) *Reports of the Secretary for Native Affairs and Chief Native Commissioners for the Year 1951*, Salisbury.

Stanning, J. (1987) 'Household grain storage and marketing in surplus and deficit communal farming areas in Zimbabwe: preliminary findings', in M. Rukuni and C.K. Eicher (eds), *Food Security for Southern Africa*, UZ/MSU Food Security Project, Department of Agricultural Economics and Extension, University of Zimbabwe, Harare, 245–91.

Sterkenburg, J.J. (1987) *Rural Development and Rural Development Policies: Cases from Africa and Asia*, Utrecht, Elinkwijk.

Stoneman, C. (1979) 'Foreign capital and the reconstruction of Zimbabwe', *Review of African Political Economy* 11, 62–83.

—— (1984) 'Problems and prospects of industrialisation in Zimbabwe', paper presented at ROAPE Conference on the World Recession and the Crisis in Africa, University of Keele.

—— (1988) 'The economy: recognising the reality', in C. Stoneman (ed.), *Zimbabwe's Prospects*, London, Macmillan, 43–62.

—— (ed.) and Cliffe, L. (1989) *Zimbabwe: Politics, Economies and Society*, London, Pinter.

Strassman, W.P. (1987) 'Home-based enterprises in cities of developing countries', *Economic Development and Cultural Change* 36 (1), 121–44.

Stren, R. (1982) 'Underdevelopment, urban squatting and the state bureaucracy: a case study of Tanzania', *Canadian Journal of Development Studies* 16 (1), 67–91.

Swilling, M. (1988) 'Taking power from below: local government restructuring and the search for community participation in C. Heymans and G. Totemeyer (eds), *Government by the People?*, Cape Town, Juta, 182–201.

Tattersfield, J.R. (1982) 'The role of research in increasing food crop potential in Zimbabwe', *Zimbabwe Science News* 16 (1), 6–10, 24.

Teedon, P. (1990) 'Contradictions and dilemmas in the provision of low-income housing: the case of Harare', in P. Amis and P. Lloyd (eds), *Housing Africa's Urban Poor*, Manchester, Manchester University Press.

—— and Drakakis-Smith, D.W. (1987) 'Urbanization and socialism in Zimbabwe: the case of low-cost urban housing', *Geoforum* 17 (2), 305–24.

Tenga, R.W. (1988) 'Land policy and law in Tanzania: an appeal for action', paper presented to the Commonwealth Association of Planners' Africa Regional Conference on Urbanization Strategies for SADCC Countries, Ardhi Institute, Dar es Salaam.

Thompson, C.B. (1984) 'Zimbabwe in Southern Africa: from dependent development to dominance or cooperation', in M. Schatzburg (ed.), *The Political Economy of Zimbabwe*, New York, Praeger, 197–217.

Tibaijuka, A.K., Maganya, E., and Naho, A. (1988) 'A study of the socio-economic effects of structural adjustment programmes: the case of Tanzania', research proposal submitted to UNICEF and SIDA.

Todes, A. and Watson, V. (1985a) *Local Government Reorganization: Government*

Proposals and Alternatives in Cape Town, Working Paper No. 33, Urban Problems Research Unit, University of Cape Town.

——— (1985b) 'Theoretical and methodological issues in the study of local government in southern Africa; local government reform in South Africa: an interpretation of aspects of the state's current proposals', *South African Geographical Journal* 67 (2), 201–11.

——— (1986a) 'Local government reform, urban crisis and development in South Africa', *Geoforum* 17, 251–66.

——— (1986b) *Local Government Reorganization: Government Proposals and Alternatives in Cape Town*, Working Paper No. 34, Urban Problems Research Units, University of Cape Town.

——— and Wilkinson, P. (1986) *Local Government Restructuring in South Africa: the Case of the Western Cape*, Conference on 'Western Cape: Roots and Realities', Centre for African Studies, Cape Town.

Tordoff, W. (1988) 'Local administration in Botswana', *Public Administration and Development* 8, 202.

Torr, L. (1987) 'Providing for the "better class native": the creation of Lamontville, 1923–1933', *South African Geographical Journal* 69, 31–46.

Underwood, G.C. (1986a) 'Zimbabwe's urban low-cost housing areas: a planner's perspective', *African Urban Quarterly* 2 (1), 24–36.

——— (1986b) 'Zimbabwe', in N. Patricios (ed.), *International Handbook of Land Use Planning*, New York, Greenwood, 185–218.

Vagale, L.R. (1987) 'Review and revision of the development plan of Lusaka, in the regional and national context', United Nations Centre for Human Settlements, Nairobi.

Weaver, J.H. and Kronemer, A. (1981) 'Tanzanian and African socialism', *World Development* 9 (9/10), 839–50.

Weiner, D. (1988) 'Agricultural transformation in Zimbabwe: lessons for South Africa after apartheid', *Geoforum* 19 (4), 479–96.

Wekwete, K. (1987a) 'Growth centre policy in Zimbabwe', Occasional Paper No. 7, Department of Rural and Urban Planning, University of Zimbabwe.

——— (1987b) 'Development of urban planning in Zimbabwe: an overview', Occasional Paper No. 5, Department of Rural and Urban Planning, University of Zimbabwe.

——— (1989) 'Urban local government finance – the case of Harare City in Zimbabwe', mimeo, Department of Rural and Urban Planning, University of Zimbabwe, Harare.

West, M. (1982) 'From pass to deportation: changing patterns of influx control in Cape Town', *African Affairs* 81, 463–77.

Western, J. (1981) *Outcast Cape Town*, London, Allen & Unwin.

——— (1985) 'Undoing the colonial city?', *Geographical Review* 75, 335–57.

Whitlow, J.R. (1980) 'Environmental constraints and population pressures in the tribal areas of Zimbabwe', *Zimbabwe Agricultural Journal* 77, 173–81.

Wilkinson, P. (1983) 'Housing', *South African Review* 1, Johannesburg, Ravan, 270–7.

Wolff, J. (1987) 'The Regional Services Council System in South Africa: a critical evaluation with particular reference to their fiscal principles', unpublished report, Urban Foundation, Johannesburg.

Wood, B. (1988) 'Trade union organisation and the working class', in C. Stoneman (ed.) *Zimbabwe's Prospects*, London, Macmillan, 284–308.

World Bank (1989) *World Development Report*, Oxford, Oxford University Press.

Yates, P. (1981) 'The prospects for socialist transition in Zimbabwe', *Review of African Political Economy* 18, 68–88.

Young, T. (1988) 'Recent development in regional government in South Africa with special reference to Regional Services Councils', mimeo, Institute of Commonwealth Studies, London.

Yudelman, M. (1964) *Africans On The Land*, Cambridge, Mass., Harvard University Press.

Zafiris, N. (1982) 'The People's Republic of Mozambique: pragmatic socialism', in P. Wiles (ed.), *The New Communist Third World*, London, Croom Helm, 114–64.

Zimbabwe, Government of (1988) *Quarterly Digest of Statistics*, Harare, Central Statistical Office.

Zinyama, L. (1982) 'Post-independence land resettlement in Zimbabwe', *Geography* 67, 149–52.

—— (1986) 'Agricultural development policies in the African farming areas of Zimbabwe', *Geography*, 71 (2), 105–15.

—— (1987) 'Gender, age and the ownership of agricultural resources in the Mhondoro and Save North communal areas of Zimbabwe', *Geographical Journal of Zimbabwe* 18, 1–14.

—— (1988a) 'Commercialisation of small-scale agriculture in Zimbabwe: some emerging patterns of spatial differentiation', *Singapore Journal of Tropical Geography* 9 (2), 151–62.

—— (1988b) 'Changes in settlement and land use patterns in a subsistence agricultural economy: a Zimbabwe case study, 1956–1984', *Erdkunde* 42 (1), 49–59.

—— (1988c) 'A comparative analysis of social and economic factors influencing agricultural change and development in the Mhondoro and Save North communal areas of Zimbabwe', D.Phil. thesis, University of Zimbabwe, Harare.

—— (1988d) 'Farmers' perceptions of the constraints against increased crop production in the subsistence communal farming sector of Zimbabwe', *Agricultural Administration and Extension* 29 (2), 97–109.

—— and Whitlow, R. (1986) 'Changing patterns of population distribution in Zimbabwe', *Geo Journal* 13 (4), 365–84.

INDEX